Integrating Crops and Livestock

Editor
Dr. Bibek Ghosh

2025

GENE-TECH BOOKS

ISBN 978-81-89729-76-9

Published by: **GENE-TECH BOOKS**®
4762-63/23, Ansari Road, Darya Ganj,
New Delhi - 110 002
Phone: 09899295259, 011-43003222
E-mail: genetechbooks@yahoo.co.in

PRINTED IN INDIA

Integrating Crops and Livestock

Preface

Integrated farming system is a system, where crops, livestock, poultry, fish etc. are raised in a farm. Integrated farming utilises cropping patterns which involve the raising of crops, animals, and or trees. A labour intensive integrated farming practice includes crops, duck, fish etc. on small-holdings, based on maximum recycling principles. Most integrated farms have more than one enterprise and may have production systems such as crop-cattle-goat-poultry-fish, crop-cattle-buffalo-poultry, cattle-goat-poultry, crop-fish, etc.

The advantages in integrated farming are: greater stability of yield over different seasons; better use of growth resources; better control of weeds, pests, and diseases; erosion control through providing continuous leaf cover over the ground surface; and benefit to small farmers of limited means. With appropriate crop combinations and proper agronomic practices, integrated farming will not only increase grain yields and ensure economic returns, but it will also make available sufficient livestock feed and maintain soil health.

This book discusses the modern trends in integrated farming and describes technologies for mixed farming systems and identifies major constraints. It is designed for students of agriculture, farmers and decision-makers concerned with agriculture and rural development.

Editor

Contents

Contents

1

Mixed Farming Systems

Mixed farming exists in many forms depending on external and internal factors. External factors are weather patterns, market prices, political stability, technological developments, etc. Internal factors relate to local soil characteristics, composition of the family and farmers' ingenuity. Farmers can decide to opt for mixed enterprises when they want to save resources by interchanging them on the farm—because these permit wider crop rotations and thus reduce dependence on chemicals, because they consider mixed systems closer to nature, or because they allow diversification for better risk management.

Mixed farming is common worldwide, in spite of a tendency in agribusiness, research and teaching towards specialised forms of farming. Obviously, mixing has both advantages and disadvantages. For example, farmers in mixed systems have to divide their attention and resources over several activities, thus leading to reduced economies of scale. Advantages include the possibility of reducing risk, spreading labour and re-utilising resources. The importance of these advantages and disadvantages differs according to the sociocultural preferences of the farmers and to the biophysical conditions as determined by rainfall, radiation, soil type and disease pressure.

There is wide variation in mixed systems. Even pastoralists practise a form of mixed farming since their livelihood depends on the management of different feed resources and animal species. At a higher level, a region can consist of individual specialised farms and service systems that together act as a mixed system. Other forms of mixed farming include cultivation of different crops on the same field, such as millet and cowpea or millet and sorghum, or several varieties of the same crop with different life cycles, which uses space more efficiently and spreads risks more uniformly.

The study of a wide variety of mixed systems at different levels is beneficial to understanding the logic of mixed systems in general. Disciplines such as ecology, economics and complex system theory have tools and concepts that can help us to understand better the mixed blessings of mixed systems. One essential point here is that the principle of mixing occurs everywhere, also in society—domestic waste such as glass, bottles or paper is also recycled.

Another point is that in mixing the different functions of plants and animals can be observed: a cereal crop produces grain and straw, a legume provides grain, organic matter, fodder and nitrogen. A third point is that it tends to be more important to look for high yield of the combination of the components rather than for the (high) yield of one component.

Mixed farms are systems that consist of different parts, which together should act as a whole. They thus need to be studied in their entirety and not as separate parts in order to understand the system and the factors that drive farmers and influence their decisions. That principle is here referred to as the "command ideotype". It may be the most important principle to achieve increased production in mixed systems, together with the awareness that crops and animals have multiple functions.

Forms of Mixed Farming

Mixed farming systems can be classified in many ways—based on land size, type of crops and animals, geographical distribution, market orientation, etc. Three major categories, in four different modes of farming, are distinguished here. The categories are:

- On-farm versus between-farm mixing
- Mixing within crops and/or animal systems
- Diversified versus integrated systems

The modes of farming refer to different degrees of availability of land, labour and inputs, ranging from plenty of land to a shortage of land. The modes are characterised by Schiere and De Wit as expansion agriculture (EXPAGR), low external input agriculture (LEIA), low external input agriculture (HEIA) and new conservation agriculture (NCA, a form of land use where shortages are overcome by more labour, more inputs and keen management).

On-farm vs. Between-farm Mixing

On farm mixing refers to mixing on the same farm, and *between-farm mixing* refers to exchanging resources between different farms. On-farm mixing occurs particularly in LEIA where individual farmers will be keen to recycle the resources they have on their own farm. Between-farm mixing occurs increasingly in HEIA systems—in countries such as the Netherlands it is used to mitigate the waste disposal problems of specialised farming.

Crop farmers use dung from animal farms. Between-farm mixing also occurs at the regional level—in the store cattle systems of the United Kingdom and the United States, animals are raised in one area to be fattened in another area where plenty of grain is available. In tropical countries also, manure may be transported from livestock farms to farmers and vegetable cropping areas where manure is in short supply.

Pastoralists from such systems in West Africa and on the Indian subcontinent also exchange cattle and crop products with crop farmers. Cultivators receive manure, labour and, less important, milk in return for cash, grain and water rights traded to pastoralists. Entrustment of livestock from crop farmers to pastoralists follows more or less the same rules. In return for taking care of the herd, herders receive either cash, or cropland, or labour for the cropland or a share of the milk and the offspring.

Mixing between nearby farms is considered here as providing the same advantages as on-farm mixing, but it should be underlined that there are important differences in terms of social organisation and transaction costs. For example, in West Africa the exchange between farms leads to tension and accidents as crop farmers start to use land that used to be pastoral only. The amount of grazing land is decreasing and dependence on the grazing of crop residues is increasing. When herders bring in the animals before the field is properly harvested serious incidents and conflicts can arise.

Mixing within Crop and Animal Systems

Mixing within crop and/or within animal systems refers to conditions where multiple cropping is practised, often over time, or where different types of animals are kept together, mostly on-farm. Both these systems occur frequently though they are not always apparent.

Within-crop mixing takes place where crop rotations are practised over and within years. For example, a farmer has a grain-legume rotation to provide the grain with nitrogen or a potato-beet-grain rotation to avoid disease in the potatoes. Plants can also be intercropped to take maximum advantage of light and moisture, to suppress weeds or prevent leaching of nutrients through the use of catch crops.

Examples of mixing between animals are found in chicken-fish pond systems where chicken dung fertilises the fish pond; in beef-pork systems where pigs eat the undigested grains from the beef cattle dung; or in mixed grazing such as cow-sheep mixes to maximise biomass utilisation or to suppress disease occurrence.

Diversified vs. Integrated Systems

Diversified systems consist of components such as crops and livestock that co-exist independently from each other. In particular, HEIA farmers can have pigs, dairy and crops as quite independent units. In this case the mixing of crops and livestock primarily serves to minimise risk and not to recycle resources.

Integration is done to recycle resources efficiently. It occurs in mixed ecological farms of temperate countries, but also in mixed, relatively low input farms of southern and southwestern Australia with grain-legume-sheep mixtures. Integration occurs most often, however, in LEIA farming systems that exist in many tropical countries where products or by-products of one component serve as a resource for the other—dung goes to the crops and straw to the animals. In this case the integration serves to make maximum use of the resources. Unfortunately, these systems tend to become more vulnerable to disturbance because mixing of resource flows makes the system internally more complex and interdependent.

In Asia, the integration of livestock, fish and crops has proved to be a sustainable system through centuries of experience. In China, for example, the integration of fishpond production with ducks, geese, chickens, sheep, cattle or pigs increased fish production by 2 to 3.9 times, while there were added ecological and economic benefits of fish utilising animal wastes. Environmentally sound integration is ensured where livestock droppings and feed waste can be poured directly into the pond to constitute feed for fish and zooplankton. Livestock manure can be used to fertilise grass or other plant growth that can also constitute feed for fish. Vegetables can be irrigated from the fishponds, and their residues and by-products can be used for feeding livestock.

Grazing of livestock under plantation trees such as rubber, oil palm or coconut is a form of crop-livestock integration that is often found in Southeast Asia. Experiments in Malaysia with cattle and goats under oil palm showed better oil palm bunch harvest and comparable results were found where goats fed under rubber trees.

In rubber and oil palm plantations in Malaysia, the integration of livestock to utilise the vegetative ground cover under the tree canopy increased overall production and saved up to 40 percent of the cost of weed control. Similarly, sheep helped to control weeds in sugar cane fields in Colombia. This suppressed the costs of herbicides, reduced the cost of weed control by half and provided additional income from meat production. This also occurs where cows graze under coconuts.

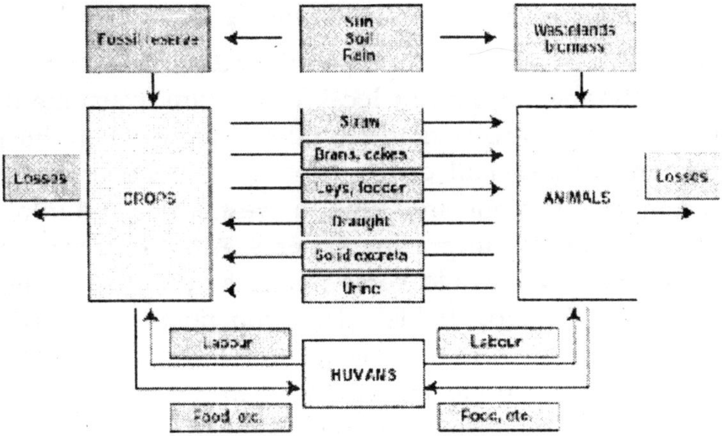

Figure 1: An outline of different resource flows in mixed crop-livestock systems

The best known type of integrated mixed farming is probably the case of mixed crop-livestock systems. Cropping in this case provides animals with fodder from grass and nitrogen-binding legumes, leys, weeds and crop residues. Animals graze under trees or on stubble, they provide draught and manure for crops, while they also serve as a savings account (Figure 1).

Mixed Crop-Livestock Systems

Even in integrated systems the exchange of resources such as dung, draught and crop residues takes place in degrees that differ among the so-called modes of farming, based on the availability of land, labour and capital respectively:

- — Expansion agriculture (EXPAGR)
- — Low external input agriculture (LEIA)

— High external input agriculture (HEIA)

— New conservation agriculture (NCA).

The EXPAGR mode occurs where land is abundant, i.e. where shortage of land or local fertility are overcome by migration or by expansion into other regions where bush and forest fallow still occur. Typical examples of mixed farming in this mode are found in West Africa and in old Asian and European grazing systems. Animals were sent out to graze and would come home to "pull the plough or fertilise the crop fields". The crop fields themselves could move elsewhere if local soil fertility declined. However, this mode is becoming more rare as land resources are exhausted throughout the world.

Mixed farming in LEIA occurs where the shortage of land can no longer be overcome by migration or use of substantial areas elsewhere for grazing. Lack of access to external inputs such as fuel, chemical fertilisers or pesticides implies that only increased use of labour and skills offers a way out. This also implies the introduction of modified practices, and the need to adjust demand according to resource availability. Dung is carried around on the farm by using more labour because a lack of soil fertility cannot be compensated by shifting to more land or by employing more livestock to "produce" more dung. In LEIA systems the latter is considered a resource but a waste product in HEIA systems. If not managed properly and if demand for food and other crops is not adjusted to the carrying capacity of the soil, this can result in mining of soils and/or collapse of the systems. Van der Pol calculated that the cotton-cereal systems in southern Mali earned 40 percent of their income by mining the soil. However, this cannot go on indefinitely and sooner or later the system will collapse if there are no changes. Some researchers think that animals, when managed correctly, can serve to fill part of the gap that exists between the output and the input of nutrients in the system, together with a proper use of chemical fertiliser.

Mixed farming in the HEIA mode is not frequently found because it implies plentiful access to resources such as external feed and fertiliser that make exchange and recycling of resources at farm level not relevant. The use of external resources can reach such high levels that the environment is affected by emissions from the crop and/ or animal production systems, ultimately leading to waste disposal problems, thus forcing HEIA into NCA.

NCA is a mode of farming where production goals are matched as closely as possible to the resource base. This approach represents a mix between HEIA and LEIA, i.e. it aims to replace the removed nutrients but it also aims to achieve keen farming and adjusted cropping and consumption patterns to suit local conditions. The use of leys is important to regenerate soils, to add nitrogen, to mobilise phosphate and to suppress weeds.

Traditional Technologies

The application of technology is not necessarily progress, but may be a response to a shortage situation. For example, in EXPAGR with ample feed available there is no incentive to plant fodder unless the existing fodder base runs out. In that case the farmers start to feed straws for lack of grass (LEIA), or they start to grow fodder based on fertiliser or legume-nitrogen (NCA). Farmers will not be convinced to vaccinate or house the animals if there is no danger of disease or predators; attention to veterinary aspects is a necessity, not a luxury. Then again, one technology may serve several goals at once and it may also carry disadvantages. Housing helps to protect the animals, to facilitate daily chores or to conserve feed. It also costs money and resources, it constrains the freedom of the animals and it may even negatively affect the health of the animals.

Similarly, there is hardly any technology that comes alone. The use of a milking machine requires the availability of skilled technicians and training of a farmer; the introduction of purebred pigs requires more elaborate housing, arrangements for supply of genetic material and availability of veterinary care. In some cases even the local food habits may have to change—for example, when slow growing pigs with much fat are replaced by fast growing pigs that have relatively lean meat and that use certain kinds of feed more efficiently.

Technology in general can be defined as a method to overcome a problem. A farmer getting tired of milking the cow will be happy if there is a labourer or a machine to do it instead. Farmers who get tired of hand weeding the crops will be happy to use a hoe or chemicals to make the job easier. Any intervention can be considered a technology—whether it is technical, management or policy related—but there is one condition. The cost of the technology should be paid by a combination of the extra return and the saved effort, a "calculation" that tends to be done differently among farmers themselves, scientists and/or policy-makers.

Several different technologies exist in relation to mixed farming and the following distinctions are made in this chapter:

- — versus management-based technology
- — Accelerating versus defusing technology
- — Exogenous versus indigenous technologies
- — Technologies for national and local problems
- — Technologies for individual farmers and for society
- — Exploitative and regenerative technologies.

As in any other classification there is always another way to distinguish between types of technology. The purpose of this classification is to show major differences.

Input- and Management-based Technologies

Input- and mechanisation-oriented technologies increase the output of a particular animal or farm by using more inputs, by using a machine to save labour, or by digging wells if there is not enough water. Management-based technologies are focused on trying to understand the farm as a combination of soil, plants or animals; the region as a combination of farms, people, mountains, underground water reservoirs, etc. Proper management can help to avoid losses where possible. For example:

- The cow when it stands up in the morning takes some time to stretch and to urinate/defecate. Knowing this, the farmer waits for fifteen minutes before letting the cow out so that the dung and urine are collected in the animal shed.
- When building the sheds, the housing can be oriented either to let the sun shine in unnecessarily and at the wrong moment, or the sun enters at times when it suits the management.
- Much effort can be spent on collection and preservation of animal manure but appropriate management is required to apply it at the appropriate moment to serve its purpose.
- Veterinarians can be called in to cure disease but proper management (hygiene, timely feeding) and a keen eye can help to prevent the need for a cure.

Accelerating and Defusing Technologies

The distinction between accelerating and defusing

technologies is difficult, but it is related to the earlier distinction between input and management oriented technologies, and to the exploitative-regenerative technologies. An example of an accelerating technology is when a cow is given a medicine that "forces" the animal to become pregnant even when the cow's body is too weak to conceive. By forcing it to conceive again the cow will collapse and even more inputs are required to get the cow in shape again.

Liming of the soil or cultivation of *Brachiaria* or cassava on poor soils can also be considered as accelerating technologies. *Brachiaria* is a fodder grass and cassava is a tuber crop for human consumption, but they have in common that they both grow well on poor soils. However, they exhaust an already poor soil even more. The same process takes place when goats or cattle are left to graze already eroded hillsides. Liming, also called marling, was a well-known form of fertilisation in Europe for many centuries. It was done, among other reasons, to release phosphorus from the soil-nutrient complex by adding calcium. Ultimately it led to soils being depleted, as evident from this example of old British farmers' wisdom of several centuries ago:

> "It brings the grounde to be starke nought, wherby the common people have a speache, that ground enriched with chalk makes a riche father and a beggerly sonne."

or in more common language:

> "Lime and lime without manure, makes both land and farmer poor."

Defusing technology as opposed to accelerating technology tends to consider the problem to a greater depth, to look for its root causes and see how the basic processes can be stalled or reversed. Defusing technologies are relevant in NCA and they tend to be approaches that fallow land, that rotate the crops rather than treating them with chemicals to enhance fertility or to combat weeds / disease, or that reduce erosion rather than living with the consequences of erosion.

By accepting natural limitations the defusing technologies would use tolerant breeds such as N'dama or Baoulé in areas with sleeping sickness, where otherwise heavy control measures would be required. When a cow is weak a diffusing management practice would be to stop milking her a few weeks sooner to help her rebuild the reserves necessary for the next lactation. Ultimately, the diffusing management approach implies attention to stress signals from the environment. It aims to reduce or adapt consumption and to restore the resource base rather than to aim beyond carrying capacity.

Indigenous and Exogenous Technologies

The distinction between indigenous and exogenous technology is again not strict but it serves to make a difference between solutions generated and collected over many generations by farmers themselves and solutions that come from outside. By stressing the existence of indigenous technology one unlocks a vast potential of local knowledge and creativity that helps to modify external technologies for local use. Farmers can often come up with cheap solutions that would not have been generated from outside. Technical solutions from outside can certainly be useful—a vaccine against disease, a new way of conserving feed, etc. Others, however, do not fit the local conditions and it may be better to use a local and cheaper solution.

Typical examples of indigenous knowledge are ethnoveterinary medicine, where women know the local herbs to cure or prevent disease in their goats, chickens or other animals, and knowledge about the best timing for ploughing or planting. Use of indigenous technologies can also be relevant for coping with variations in soils, herding rights, local business people, etc. Indigenous knowledge is no "cure for all", e.g. it has no or almost no ways to cope with infectious diseases and occurrence of new disorders that come with development. Still, the use of local knowledge for local problems has been shown to improve the uptake of development programmes substantially.

Technologies for National and/or Local Problems

Some problems in animal production are felt more at national than at local farmers' level. Rinderpest, foot-and-mouth disease (FMD), classical swine fever, etc. cause damage for the farmers, but they mostly threaten the export licences of countries. In other words, many farmers will not feel motivated to do things that are not in their own direct interest. Large-scale programmes will have to take that into account. A more dramatic example is when governments want food to be cheap and plentiful for urban populations. This may require the production of cereal varieties that are not liked by the farmers themselves. It may even force governments to work against local traditions that, for example, consider milk as a gift from the gods that is not to be traded for profit. Improper handling of such tensions will not lead to effective development.

Technologies for Individual Farmers

Cooperatives are useful forms of farmers' organisations for development, for example to supply inputs, to work on credits and savings, or to work on a watershed programme. However, not all technologies are useful for collective action. For example, a vaccination campaign almost certainly needs the organisation of groups, whereas the treatment of a broken leg or a difficult birth in an animal is based on individual interests. Farmers' study groups are collective action but farmers' investment decisions and cropping/livestock keeping strategies are the decisions of the farmers and/or their extended families themselves.

Exploitative and Regenerative Technologies

This distinction treats the time horizon of the farmers and of farming. Planting of Brachiaria leads to rich yields but also to more exhausted soils. The planting of legumes or investments in soil regeneration give lower short-term yield but ensure a livelihood for the future generation. In particular, the mixed farming of the NCA mode puts heavy emphasis on regeneration by involving technologies such as the use of leys, recycling of nutrients, mutual adjustment of cropping/livestock components and erosion control through fodder crops.

Mixed Farming and Communal Ideotype

One particular issue in the suitability of technologies for mixed systems is in the concept of the "communal" ideotype. This term was coined by Donald for wheat breeding and it emphasizes that the yield of an individual plant needs to be made subject to the yield of the entire plot. More practically, no farmer is interested in high individual plant yield if it does not lead to the increase of total farm yields. The principle of the communal ideotype and choice of technologies for the design and choice of technologies in mixed crop-livestock farms can be illustrated in many ways:

— In fodder crops it has been shown that napier grass and leucaena can be grown alone or in combination, particularly for NCA conditions. Growing of only one crop gives higher yields per crop but a lower yield from the combination (Table 1).

Table 1. Yields of napier (Pennisetum purpureum) and leucaena (Leucaena leucocephala) in pure and mixed stands

	Yield (tonnes/ha/yr)		
	Napier	*Leucaena*	*Total*
Napier alone	12	-	12
Leucaena alone	-	8	8
Napier + leucaena	10	6	16

— A study for LEIA conditions in India with different crop-animal combinations showed that the use of cows with individual productions of 10 litres/animal/day led to higher total farm outputs than those producing 16 litres/animal/day (Table 2).

Table 2. Optimum crop combinations, herd size and production at different individual cow productions with or without treatment of stover, when the farmer also has access to a small fixed area of good quality fodder

Individual production	*Total production (litres/day/system)*	*Herd size (cows/farm)*	*Cotton(ha)*	*Total income from milk and crop (litres/day/cow) (Rs/day/farm)*
2.0	5.1	2.5	0	22.2
6.0	9.5	1.6	0	35.4
10.0	10.6	1.1	0.4	39.1
16.0	6.6	0.4	1.0	27.6

The high- producing dairy cows cannot be fed with fibrous crop residues such as straw, i.e. the goal of high milk yield per cow would reduce total farm income, thus losing an opportunity to use these resources.

— A study of crop-livestock combinations in Kenya shows an added issue, i.e. the optimum combination of animals and crops. The large farmer benefits more from grade cows (high yielding) and the small farmer is better off with crossbreeds.

Field observations and common sense support the point that farmers adjust parts of their farms to achieve larger overall output. In many tropical and temperate contexts they use crossbred cows or medium-producing animals rather than purebreds. In drought-prone areas farmers choose grain crops that also yield good straw/fodder, even at the expense of some grain yield.

Integrating Crops and Livestock

Most livestock products in mixed farming systems are derived from animals that are fed on local resources such as pasture, crop residues, fodder trees and shrubs. Another feed resource consists of by-products from village industries such as oil extraction, grain milling and cane crushing. The exchange of these resources saves money and reduces waste by recycling products within the farming system. They create employment and they can contribute to soil texture and fertility while being an economic incentive for rearing multipurpose animals.

As in the discussion of the other topics, relevance of technologies depends on the mode of farming. For example, in EXPAGR the animals, crops and soils are not strongly linked. Animals tend to graze elsewhere, they are a savings account and they may bring home some dung and young stock to pull the plough. However, in NCA the crops and livestock are integrated, as in LEIA, and they both serve the same goals—to keep the farming sustainable and productive. Still, under poverty conditions in LEIA, there is a tendency for good-quality feed residues to be sold off the farm, thus leaving the animals with mainly fibrous feeds. Therefore animals in LEIA tend to be a way to convert fibrous feeds, crop residues, poor-quality grazing land and weeds into draught, dung and urine, often at the expense of the land. In HEIA there is hardly any attention to interactions among soil, plants and animals because limiting nutrients can be bought from outside and waste is ignored until it becomes a serious problem. HEIA exists not only in temperate and rich countries, but also in tropical and poor countries.

Crop Residue and Soil Management

Livestock can enhance farm output by intensifying nutrient and energy cycles, particularly in LEIA and NCA. In LEIA this mainly implies a closer recycling of nutrients in straws (crop residues and fodder), dung and urine. In NCA there is scope to include leys, anti- erosion bunds and, for example, adjusted cropping patterns. NCA also has the added advantage that it can both import some nutrients in fertiliser and feed while it aims to keep the better-quality crop residues such as brans and cakes on the farm.

Crop residues can be used for animal feed as well as for soil fertility and can also be sold. The farmers have to choose how to use these materials and, whatever they are used for, the choice needs to be accompanied with keen management to make the best of these resources. The fibrous crop residues have a potential in soil conservation through mulching but even the oilseed cakes can be used to improve soils.

Cycling of Nutrients

The use of dung and urine excreted by cattle, pigs, chickens and other animals for crops is traditionally raised as an argument in favour of keeping livestock. Dung and urine contain several nutrients such as nitrogen, phosphorus and potassium, and the solid fraction contains organic matter that is important to maintain soil structure and fertility.

Occasionally it is even believed that dung and urine have medicinal qualities, but more commonly the dung is used as fuel, while it can also be used to plaster walls when mixed with clay. The belief in the medical value may have some justification, e.g. through the occurrence of antibiotics in dung and through the presence of lignin derivatives in dung that help to make it sticky or even to give it bactericidal properties.

The use of dung and urine for soil fertility is most cited. It is particularly important in LEIA and NCA, but unimportant in EXPAGR. The term dung "production" is misleading because animals excrete only nutrients that they first have to eat. Animals can carry nutrients from one place to another, they can increase the rate of their turnover or they can concentrate the nutrients from the outfield to the infield, i.e. they concentrate the fertility of a large land area on relatively small cropping areas around the villages. The "production" of dung in HEIA is a disadvantage rather than a blessing and elaborate systems are designed to dispose of dung rather than to use it.

Organic matter comes from manure, household waste and compost, or from incorporating weeds into the soil. The tasks involved in producing and using organic fertiliser are divided among different members of the household. Young people or labourers tend to collect the crop residues, to dig compost pits when needed and to transport the organic fertiliser to the field while the head of the family is responsible for managing its production.

Women make organic fertiliser from household waste and keep some of it back for use on their own small plots of land. The major nutrients in fertiliser and dung are needed to make plants grow and produce:

— Nitrogen is essential for growth, for making the stems, leaves and roots; it is an important component of proteins.

— Phosphorus is needed for the plants to flower, to bear fruits and to develop strong roots.

— Potassium is important for tuber and fruit enlargement while it helps to maintain the healthy activity of all plant tissues.

Manure and Nutrient Cycling

The amount and quality of urine and dung produced depends on the type of animal, its size and the type of feed as well as on the management of the farmer. One way to calculate the amount of manure produced is:

— One animal of 250 kg live weight has a feed intake that averages 2.5 percent dry matter of its live weight. It therefore consumes 250*365*0.025 = 2 280 kg of dry matter. With an average digestibility of 55 percent, the animal will produce 0.45*2 280 = 1 026 kg of dung every year.

— Small ruminants weighing 25 kg and feeding 3.2 per cent daily on average of their live weight, consume 25*365*0.032 = 292 kg dry matter. If their average digestibility of the feed is estimated at 60 percent it can be calculated that one small ruminant produces some 117 kg of dry matter faeces per year.

After the harvest a part of the crop residues tends to be freely accessible to all livestock including those from outside the village. The residues consist of straw that is sometimes left in the field or carried home and they contain stubble and ratoon which can be grazed. Another part of the residue is fallen grain—in Asia this is used by ducks. Weeds can also be considered as crop residues and in wetter areas or seasons they can form an important feed resource. Sometimes they are cut and carried home during the cultivation period when access to grazing grounds is difficult for animals.

Table 3 gives an indication of the concentration of the main nutrients in the manure of cattle and small ruminants and other sources of organic fertiliser. Dry matter content of manure also varies widely; in cows on lush pasture it can be less than 15 percent but in sheep on dry forage it can be higher than 50 percent.

Table 3. Nutrient contents of manure and other organic fertilisers (%)

Organic fertiliser	*N*	*P*	*K*	*Dry matter*
Cattle				
Fresh manure	1.4-2.8	0.5-1.01	0.5-0.6	15-25
Kraal (litter)	0.5-2.3	0.22-0.81	0.77-5.44	40-60
Kraal (no litter)	1.5-2.5	0.2-0.6	1.5-2.0	30-50
Goat and sheep				
Fresh manure	2.2-3.7	0.25-1.87	0.88-1.25	50-70
Fresh green manure	2.0-4.3	0.1-0.3	1.0-3.4	
Compost	0.3-0.9	0.07-0.17	0.14-1.3	
Household waste	0.2-0.9	0.05-0.5	0.1-2.1	
Ash from farm cooking	0.2-0.6	0.1-0.6	1.1-2.7	

The amount and proportion of nitrogen excreted depend on animal diet. The urine and solid dung of animals fed highly digestible diets with a lot of protein contain much more nitrogen and, therefore, are more susceptible to nitrogen losses than excreta from diets containing greater amounts of roughage. Much of the urine nitrogen is lost via ammonia volatilisation. Where animal management tends towards increased stall-feeding, the composting of fresh manure will have to play a greater role in minimising nutrient losses. Pits or heaps that capture feed refusals, manure and urine and household waste need to be designed to minimise nutrient losses. Low-cost appropriate implements to spread the compost at the appropriate time over large cultivated areas such as the Sahel are also needed.

In various systems draught oxen, donkeys, horses and small ruminants are kept overnight in pens in the compound throughout the year, and the manure they produce is transported to the fields during the dry season. Droppings of donkeys, sheep and goats are often added to the heap of household waste. However, droppings from small ruminants are sometimes used separately, for example on infields to manure certain spots in the millet crop.

Manure from small ruminants takes longer to have an effect on crop yields than cow dung, but once it has started the effect lasts for several years. As this dung is kept the nitrogen losses can be reduced by mixing with low quality biomass, i.e. with straws, leftover feeds, dry leaves, etc. The greener leaves still contain much nitrogen themselves and they are less capable of capturing surplus nitrogen from urine and dung.

Household Waste

Household waste consists of partially decomposed waste from a variety of origins. It plays no part in fertility management where land is abundant. However, farmers are prepared to work hard to retain nutrients on their soil when land for cultivation becomes scarce and when food demands increase (LEIA and NCA). For example, farmers in Dilaba, Mali, said they started applying household waste in the 1970s after the drought, combining it with cow dung and the droppings of small ruminants. The system is now widely used in Dilaba and it is also starting to gain ground in the surrounding areas. The amount applied by different farmers depends largely on how much they are able to transport and on how much they can collect.

Composting is a process to break down the organic materials to make the nutrients in the biomass accessible to plants. It also helps to make a fine, easy-to-handle material and, if well done, the composting kills weed seeds. Compost is normally produced from a mixture of all kinds of organic waste, including crop residues and household waste. The technology is very ancient and there are many different ways to compost organic materials. The most common method uses a moist heap with enough nitrogen for the breakdown of the organic material that is regularly turned to allow air and oxygen to enter the heap.

Nutrient Losses

Different kinds of nutrient losses can occur during crop cultivation and post harvest:

— Stray animals can damage the standing crop during crop cultivation and post-harvest period, in terms of losses of nitrogen, phosphorus and potassium and in terms of feeding value.

— Straw stubble that is left on the field can be trampled or soiled and incompletely eaten: careful harvest could save valuable nutrients.

— Leaving the crop residues in the field causes a loss of nutritive quality and sometimes leaching of nutrients through rains and degradation processes that involve fungi.

— Lush regrowth, in some cases after sudden rain, can be poisonous, such as the formation of prussic acid in sorghum ratoon.

— When the soil is bare for long periods nutrients can be lost due to erosion (wind and water).

Animals leave their droppings while they graze the stubble, so even farmers without livestock will receive some on their fields. However, the quality of the dung is likely to deteriorate if left lying on the field unprotected for up to six months, even when it is stacked and stored in the open.

Grazing animals do not "produce" nutrients in dung and urine; actually they remove more biomass and nutrients from cropland than they return in the form of manure, as shown by studies in some West African countries (Table 4).

Table 4. Cereal stover removals and manure returns during crop residue grazing in West Africa

Location	*Total (kg/ha)*	*Dry (kg/ha)*	*Matter (kg/ha)*	*Nitrogen*	*Phosphorus*
			Stover removal		
Nigeria	2 470			24.5	3.9
Burkina Faso	1 570			14.8	4.0
Niger	2 135			16.8	1.8
			Manure returns		
Nigeria	27-272			0.3-1.7	0.1-0.3
Burkina Faso	600-1 600			7.5-20.0	1.5-4.0

Activities to prevent or reduce nutrient losses are:

— Penning of animals with or without bedding.

— Storage of crop residues.

— Mulching of the soil with residues unfit for animal feed.

— Breeding or managing for crop varieties with better or more straws or for cultivars that have more ratoon.

— Covering of soil with so-called catch crops that retain leached nutrients to build their own organic matter that can be used for animal feed or mulch.

— Harvesting of crop residues before they overmature, the so-called stripping.

Cropping Patterns

The survival of farmers in mixed systems depends to a large extent on a proper adjustment of soils, plants and animals. Techniques suggested in this section are :

— better use of fallow;

— use of leys and catch crops.

These techniques can help in several ways either to reduce erosion, to add nitrogen, to enhance phosphorus flow rates, to pick up leached nutrients or to provide feed.

Fallow

Farmers in most EXPAGR regions tend to depend on fallow as the technique for restoring soil fertility. This is possible because there is plenty of agricultural land available and entire fields can be left fallow for long periods of ten years or more. The system resembles shifting cultivation and the location where crops are cultivated changes every couple of years. However, over the years, with increasing population and changing technology, the fallow period has become shorter and, in areas with a dense population, it has disappeared almost entirely.

Leys and Catch Crops

A ley is a fallow that is planted with crops such as grass or legumes to regenerate the soil more rapidly. The grass is used as fodder if animals are present on the farm, but similar systems are used when only mixing between crops takes place. In Mali, for example, farmers grow millet and cowpea on the same infield year after year. When they find that the fertility of part of the plot has fallen significantly, they change the association to pure stands of groundnuts or Bambara groundnuts. The following year they apply organic fertiliser and then start cultivating millet/cowpea again. Such a strategy is widely used in the Sahel.

In other words, a ley is a crop rotation that includes animals. To some extent it is also the system used in the Gangetic plains and in the Nile delta, where berseem and/or mustard are rotation crops used exclusively for livestock feeding, either on-farm or off-farm. The system is being tried in northern Thailand, and is also practised in the Krishna Valley of eastern India where a legume is the main rotation crop. However, use of legumes is also common in the United States, in western Europe and in the so-called Mediterranean systems that even occur in Australia.

Sustainable Land Use

Integration of crops and livestock is often considered as a step towards sustainable agricultural production because of the associated intensified organic matter and nutrient cycling. Residues of the different crops represent the main on-farm source of organic matter and nutrients. This combines well with the presence of livestock since animals play a vital role as capital assets for security and as a means of saving, for cash income and in nutrient flows. Management of crop residues in such regions is closely related to their utilisation in animal feeding.

An advantage of the integration of livestock and crops is the added value derived from crop residues in terms of animal products and income. However, to maintain the system in the long term it is necessary that nutrients from external sources are added. In the past, this was covered to some extent through fallowing and manuring contracts with pastoralists, but the growing demand for feed by the increasing herd, the shrinking area of cropland per capita and declining crop yields dictate the scope for improvement through integration of crops and livestock. The extra input must come from inorganic fertiliser as well as from concentrates or both.

Often, the price ratios of fertiliser and grains are not conducive to the utilisation of fertilisers, and development of institutional and physical infrastructure for cost-effective use of fertilisers and concentrate is required to trigger sustainable land use in the region. Use of concentrate can be remunerative for farmers but credit facilities may stimulate intensification of livestock production and thereby increase availability of nutrients for crop land.

Energy, Biomas and Nutrients

Animals can also play a role in the provision of energy. Sometimes this role is very negative where livestock keeping contributes to deforestation in large parts of South America, Asia and Africa, but it can also be positive, such as by transforming plant energy into useful work or by providing dung used for fuel through dungcakes or biogas to replace charcoal.

Wood and charcoal will con-tinue for a long time to provide energy for household cooking and for rural industries but increasing energy demands and increasing pressure on natural resources call for measures such as a more efficient use of biomass, improved natural resource management, fuel switching from, for example, charcoal to kerosene or liquefied petroleum gas, and for the development of alternatives such as biogas. The principal uses of biogas are for household energy, such as cooking and lighting, although larger installations can produce sufficient gas to fuel engines, for example, for powering mills and water pumps.

Despite the merits of biogas technology, bio digesters have only been widespread in India and in China. In these countries it is common to find biodigesters even in remote villages provided there is enough water. For example, an ambitious programme in Nepal is accelerating market development of small biodigesters. In many countries, some practical experience has now been gained with the dissemination of biodigesters but larger-scale dissemination has not yet become popular, despite the assumed economic, environmental and social benefits of the technology.

The main reasons seem to be that the multiple benefits of biodigesters must be acknowledged by the end-users to convince them to invest in the technology, and that a market-driven institutional infrastructure should be in place to facilitate large-scale dissemination of the technology. Much work has been done on mixed crop-livestock systems including the use of biogas, for example in the well-known CIPAV project in Colombia where pigs, sheep, sugar cane and biogas are just some of the components studied.

Three designs can be distinguished among the small-scale and low-cost biodigesters:

- the floating-drum, also known as the "Indian design";
- the fixed-dome, known as the "Chinese design"; and
- the flexible-bag digester.

A well-constructed fixed-dome biodigester has the longest lifetime: 20 years and more. The floating-drum digester can have a comparable lifetime, but the recurrent costs are higher as the steel drums have to be replaced every five to ten years due to corrosion. The lifetime of the flexible-bag digester is hard to indicate, as some do not last more than a couple of months, while others function for several years.

Complex microbiological processes take place within a biodigester. In a common farm-type digester, the temperature ideally varies between 30 and 40 degrees Celsius. The process can be divided into three steps. During the hydrolysis phase, the complex molecules in the feedstock are broken down into smaller molecules. During the second phase, volatile fatty acids are formed. Finally, during the methanogenic phase, other bacteria use the acids to produce methane. The methanogenic bacteria can only do their work under anaerobic conditions (without oxygen), i.e. it is important for a biogas plant to be built of leak-tight materials. In practice, the three phases of the process take place simultaneously in a digester where the different bacteria work together in a symbiotic relationship.

Many different types of organic matter can be used to "feed" a biodigester, but to allow the bacteria to do their work the organic matter should be accessible to them. This means that pretreatment is sometimes necessary, e.g. through chopping and/or composting of crop wastes. The easiest feedstock to use is cattle dung, as it already contains the right bacteria and the vegetable matter has been broken down by passing through the guts (and teeth) of the cow.

Human excrement and manure from chickens and pigs are also useful, but they do not contain the right bacteria and they need a starter in the form of, for example, some slurry from a working biodigester. Neither the feedstock nor the water should contain toxins such as antibiotics, detergents and disinfectants. Toxins cause the digester to "go sour", causing it to give off an unpleasant smell of acids and, therefore, the dung of cows that receive antibiotics should not be fed into the digester. Cultural aspects might influence the selection of feedstock. For example, Hindus in India usually accept cow dung for use in biodigesters, while pig dung and human excrement are often not accepted, particularly in Moslem regions.

Initially, biodigesters were promoted only for their energy production. Once again, the thinking about a technology in a mixed system focused on only one aspect. Unfortunately little attention has been paid to the excellent properties of the slurry as an organic fertiliser of crops and trees. In other words, slurry has the potential of supporting crop production and, in addition, the process kills or substantially reduces the amount of pathogenic germs and seeds of weeds present in the feedstock.

Another advantage is that nutrients that are already present in the feedstock, such as nitrogen, potassium and phosphorus, are made available to crops more quickly and efficiently, i.e. the biodigester affects the nutrient dynamics of the system. Further, it reduces the odours given off by manure and compost. Emissions of greenhouse gases are also reduced, as the biodigester produces renewable energy and because fossil fuel based chemical fertiliser is replaced by an organic fertiliser. A major disadvantage of the slurry is that it is liquid, with a dry matter content of less than 10 percent. Transport to the fields is therefore often complicated and the use of the slurry is usually limited to fields near the digester. Larger installations in industrialised countries separate the slurry into a liquid and a solid fraction; the solid fraction can be applied on fields at a distance from the digester.

Farm Management

A proper mix of indigenous and exogenous knowledge is likely to be effective in development of mixed systems—just as it is necessary to ally properly the interests of men, women, production and sales, etc. It will be seen that all these issues reflect the concept of the communal ideotype at farm, regional or even national levels and among disciplines.

Communal Ideotype

The essence of the communal ideotype is that parts of a farm need to be mutually adjusted to ensure maximum output of the total. This adjustment should take place not only between parts in space (i.e. the goat has to be tethered to avoid crops being damaged), but also in time (i.e. the yield objective of the present crop may not be at the expense of future yields; that would be a case of soil mining). For example:

— High-yielding cows cannot make use of fibrous crop residues; i.e. higher output of the ecosystem can be at the expense of the output from the whole farm.

— A high production of an animal in one year can exhaust the animal in such a way that productivity can be adversely affected over several years.

— Farmers reject new cereal varieties because they do not yield enough straw or straw of a good enough quality for animal feed.

— The excessive use and/or improper application of fertilisers may help to increase the yield of a fodder crop at the expense of the health of the ecosystem, i.e. ditches become polluted or lakes are overgrown with weeds.

All these issues are typical in mixed systems but, as previously mentioned, they also occur elsewhere in society, thus making the work in mixed crop-livestock systems an ideal model for other sectors of society.

At village or regional level there are also examples of development based on involvement and mutual adjustment of all partners.

Landcare and Integrated Catchment Management, Australia

Landcare and Integrated Catchment Management (ICM) are two community participative organisations with a diverse membership in rural areas that includes primary producers, nature conservationists, government and industry. Land care in this sense is not a typical case of mixed crop-livestock farming, but it does represent a practical example of how issues in mixed farming systems translate into issues in development of society in general. Both Landcare and ICM involve all members of the community in the decision-making process for the management of catchments. Both these organisations mould together private use of land for agriculture with a philosophy that all land is held in trust for the use of future generations of human and other life. They do this by planning for long-term, sustainable use of land at a catchment level and use public and private funds to implement the plans at a property level, the epitome of the communal ideotype in space and time.

Members of the organisations work together, and work with government bodies, to share information about best practices and to make incremental steps towards sustainable land use. Nature conservation groups are part of both organisations and they contribute to the thinking about what is current best practice for the needs of wildlife. They also help to implement the plans using both public funds and private donations of cash or kind. Landcare and ICM have developed into strong organisations that combine the interests of different system levels in space and time.

The Sukhomajri Project in India

Sukhomajri lies in the hills, not far from Chandigarh in northwest India. The hills, once heavily wooded, have suffered from increasing human and livestock populations, leading to over-grazing and severe erosion. Barely 5 percent of the uplands had vegetation cover in the 1970s, and erosion rates of 150-200 tonnes/ha were not uncommon. In 1958 a dam was constructed, creating a lake to serve the city of Chandigarh, but by 1974 over 60 percent of the lake was filled with sediment. To protect the lake, the authorities first tried to persuade the gujjar herdsmen of the village to stop using the hills for their cattle and goats, but with little result. The breakthrough came when it was decided to build a small earthen dam in the hills to provide water for the village itself and then to stabilise the catchment of this dam. The stored water was used to irrigate nearby fields and farmers were provided with subsidised seed and fertiliser.

Yields were greatly increased but farmers who did not benefit continued to use the hills for grazing. It was then that the villagers collectively proposed that more small dams should be created to extend the irrigation system. They also suggested the creation of a water users' society, based on the principle of equity, to manage the water. A "coupon" system was introduced and families with little or no land could thus sell their water rights or use the water to sharecrop on land belonging to others who were short of water. Any member whose livestock was found grazing in the hills lost his or her rights to the water. The society was given responsibility for maintaining the dams and their catchments, for distributing the water, and for maintaining records.

From then on the village began to develop rapidly. The villagers sold off their goats and replaced them with high-yielding buffaloes to provide milk for the growing towns nearby. The buffaloes were stall-fed, using the rapidly growing fodder grasses in the hills. These first steps immediately brought benefits and helped enlist villagers for further experiments in improvement.

For instance, the farmers began to experiment with green manures, which had not been tried before, and they developed new technologies based on velvet bean and lablab bean. As maize yields increased and subsistence was assured, the farmers began to turn to vegetable growing. The project is on a self-help basis, with no subsidies provided. By 1990 the farmers, with their own labour, had constructed some 300 km of erosion works. Out-migration has largely ceased and the landless in the area are benefiting from an increase in the daily wage. Many landless have begun to establish rights to land that previously they had considered useless but which under the new technologies is proving productive.

SIWAA, the dry forest, Mali

Natural resources are in short supply in the Koutiala area, one of the oldest cotton production areas of Mali. Villagers complain about townspeople cutting firewood and grazing herds in the bush. In an attempt to regulate the use of the natural resource base a pilot project with six villages has been set up. They have been collaborating with other users of the natural resources and with research, extension and local authorities to manage their territory.

Over the years plans have been put to work to control grazing, cutting of firewood, livestock feed and other uses of the resources. The plans included intensification of agriculture, integration of agriculture and livestock, fodder crop production, erosion control, establishment of tree nurseries, reforestation, and adaptation of improved cooking stoves. Respect of traditional laws, for example on when and how to exploit certain tree species, has been included in the protocol. The whole approach has been based on participatory methods.

Chosen villagers represented the villages and they gave feedback to the village on results reached or measures proposed to control their environment. It took a long time to get the different stakeholders together. The process was slow, with conflicting interests, but in the end an agreement was reached, the protocol was tested and is now being implemented. The cutting of firewood is regulated. There are rules for grazing, rotational grazing and bush fires, and cutting heights of trees are fixed to enable regrowth.

Role of Community Forestry

Farmers in the Hindu Kush-Himalayas had mixed reactions regarding community forestry. Some felt that it had a positive impact on fodder supply because of natural growth as a result of protection, but others thought that controlled harvesting of natural grass and leaves would decrease the availability of fodder resources. A minimum fee could be paid on a monthly basis to acquire fodder grass from community owned lands, but poor farmers did not find this acceptable. With the increasing scarcity of fodder, farmers searched for alternatives.

A greater number of farmers addressed fodder scarcity through use of their own lands. In some villages, about 30-40 percent of the land-rich farmers have private land, Kharbaris (marginal sloping land), on which grass and fodder trees are grown more intensively. Increased fodder shortages and the high cost of feed provided incentives for the increased cultivation of trees on farms. There has also been an increase of growing leguminous fodder trees on the bunds of cultivated terraces. During the winter months, maize stalks and badhar (Artocarpus lakoocha) are fed to livestock.

Farmers also save dry fodder (rice and wheat straw, maize, stover, etc.) during this period to feed large ruminants but, as the amount available is not sufficient to feed them all winter, in some villages farmers have adapted to winter fodder shortages by cultivating exotic oats, sometimes cultivated after the rice harvest, to replace wheat. Also, some resource-poor farmers have responded to the lack of fodder from their private land by making arrangements to harness fodder from the private land of resource-rich farmers.

Policy Measures

Not only farmers and their communities have to change their attitudes, also the people and institutions at government level may have to collaborate in adjusting their priorities to make livestock development and mixed farming a success. Typical cases range from the privatisation of veterinary care to the reorientation of research and extension programmes.

Veterinary Services

Traditionally the tasks of veterinary services in developing countries include active veterinary care: vaccinations, health care and marketing of veterinary and feed products. However, state-organised veterinary services have been increasingly incapable of responding to the growing demand for veterinary care because of budget restrictions, inefficient services and inefficient handling of revolving funds. Therefore, the privatisation of veterinarian services was stimulated to continue offering animal health care after the Structural Adjustment Programs of the World Bank and IMF.

Multilateral donors, such as the European Union, support private veterinarians through credit schemes and trade guarantees and the functioning of veterinary care on an open market is expected to increase availability and to improve the quality of animal health care. Better adjustment of the services offered and services needed is expected from privatisation, though this may work better for one situation than for another. The prevention of diseases of national importance may be less effectively tackled if farmers determine demand for services only.

Privatisation has indeed been successful to some extent. However, the private veterinarian hardly covers remote areas because costs of transport are prohibitive and demand for services is low. In many African countries the adjustment process is progressing slowly, but the privatisation of vaccination campaigns against bovine pest and contagious bovine pleuropneumonia has been quite successful in Chad.

The trade of veterinary products has been successfully liberalised in Mali, but at first the availability of private agents restricted the range of vaccination programmes and prevention against trypanosomiasis. The success of privatisation also depends on the organisational capacity of the veterinarians and their willingness to work in an insecure financial situation. Donors have stimulated governments to change the legal framework to facilitate the work of these veterinarians. However, the application of this framework depends on the veterinary services. Veterinarians have professional associations that can put pressure on government to either activate or discourage the new legal frameworks and adjustment processes.

Privatisation of the veterinary services in the Compagnie malienne pour le développement des textiles (CMDT) in the region of southern Mali has given mixed results. In some parts of south Mali, such as Kadiolo, the introduction of cotton and the related animal traction on the farm has been relatively recent.

Veterinary services such as vaccination and deworming of oxen were carried out twice a year on the initiative of the cotton company (CMDT). Farmers were not fully informed about the usefulness of these programmes. Sometimes costs were deducted from the cotton yield without involving farmers, who were not likely to call upon the veterinarians themselves because they were unaware of the advantages of preventive treatment. At the same time, private veterinarians were not stationed in these regions, as the demand for their services was low. In these cases privatisation had negative consequences because some farmers lost their only oxen that were acquired on credit.

In 1995 and 1996 when privatisation was under way and when state agents and the CMDT were prohibited to intervene, an outburst of pasteurellosis occurred, while trypanosomiasis caused mortality in cattle. The cotton production of these farmers decreased because of the poor health of their oxen, and so did their capacity to reimburse the credit they had received for cotton farming and the animal traction equipment.

In regions such as Koutiala, where introduction of animal traction started much earlier, farmers learned about the advantages of vaccination by experience as they invested their savings from cotton production in cattle. They were used to calling upon veterinary services for their animals. By experience they learned that only vaccination could prevent cattle from getting bovine pest, pasteurellosis, anthrax (endemic in the region) and contagious bovine pleuropneumonia. After privatisation those farmers did not hesitate to bring their oxen for vaccination to private veterinarians that were already stationed and equipped with drugstores.

Research and Education

Much of the modern teaching and research tends to be focused on single commodities and disciplines. This monofocused approach on either breed, or feed, or health, or milk, or eggs, etc. has particularly hampered the development of mixed systems where animals perform several functions and where different aspects of animal production are to be mutually adjusted. Fortunately change is imminent and more research is now undertaken on issues of mixed farming worldwide. This has led to new insights and approaches to teaching. For example cereal varieties are now tested for grain and straw characteristics as well as for ratooning capacity.

Successful Mixed Systems

Sustainable systems are those that have existed for a considerable time. However, it must be recognised that no system can exist forever, i.e. any system becomes unsustainable sooner or later. All systems have to adjust continuously, particularly when the relationship between the system and its resource base becomes seriously out of balance. An example of a system and its resource base is the case of the infield/outfield ratio in EXPAGR or HEIA, or the ratio between management requirement and management capacity when a farming system shifts from HEIA into NCA. The following cases are illustrative of previously sustainable systems. They give clues for future development and for the choices in policy setting or research/extension priorities.

Outfield/infield System

A mixed system that uses animals to graze "outfields" distant from the village in order to concentrate all or part of the excreted plant nutrients on the "infields" near the farmhouse is called an outfield/infield system. It is a form of mixing in the EXPAGR mode where land is abundant. The system exists (and has existed) in many forms. It can employ animals that are only stocked on crop fields at night while being grazed during the daytime, it can be intensified by constructing special stables in which straw from the crop land, litter from the forest or even topsoil from the outfield are used to conserve the nutrients of the dung and urine from the animals. As recycling on the farm increases, the system slowly moves into the LEIA or NCA mode. The system has existed throughout the world for many centuries, in pre-medieval Europe and in Russia until the early twentieth century, and it is still found today in many countries in the tropics. Improvements include the better conservation of nutrients in the stable, the better application of dung on the field (more timely and better localised), or the introduction of legumes for nitrogen supply.

Kano Close-settled Zone

The Kano close-settled zone in the Soedano-Sahelian region of Nigeria has been the site of an intensive mixed farming system of the NCA mode where for the last 30 years all available land has been under annual cultivation. Small ruminants consume crop residues, in particular those of groundnuts and cowpeas, which are good quality fodder. The nitrogen in the residues of leguminous crops is conserved in the manure, which is transported with compound waste back to farmers' fields for use as fertiliser.

Legume grains are sold, earning cash, which farmers may use to purchase inorganic fertiliser if they wish or other goods. Nutrients are added to the system when harmattan dust (a wind from the north carrying sand from the Sahara and North-Sahel) is deposited on farmers' fields during the dry season. The system allows farmers to manage an efficient nutrient recycling system centred on small ruminants and based on high labour inputs by farmers. They must keep animals tethered within the compound during the rainy season, collect crop residues and weeds for fodder, and transport the manure back to the fields.

Several elements have been important in the development of this farming system:

— The high population density provides labour and a source of agricultural innovation combined with a remarkably high density of small ruminants and donkeys.

- — Land tenure is organised such that farmers have usufruct rights over the land they farm.
- — The settlements are dispersed and farmers live relatively close to their fields, so as to minimise the time needed to reach the fields and to transport crop residues and manure between fields and the compound.
- — The farmers use the organic fertiliser on all their fields.
- — The investment in oxen, ploughs and cultivators by a few farmers has been viable because they can hire out the ploughing and cultivation service to other farmers.

In addition, the farmers engage in crop, livestock and tree production to diversify their income. Many also practise non-farming economic activities, especially during the long dry season. This diversity allows them to cope with risks, whether environmental (e.g. drought) or economic (e.g. price fluctuations). The dependency on external inputs is low, permitting farmers to be independent of the fluctuations in the economy. Farmers use little inorganic fertiliser, and they maintain seed lines of favoured indigenous cultivars, rather than using the seed of commercially supplied high-yielding varieties. Hence, despite being one of the most densely populated areas in semi-arid West Africa, the system is both productive and sustainable—an example of farming in the NCA mode. The increasing population densities and labour availability are for the time being essential for a process of agricultural intensification, such as the increasing use of crop residues to feed livestock, and the use of farmyard manure on the fields.

Machakos Case of Kenya

A widely quoted example of population pressure and its impact on development and environment is that of the Machakos district in Kenya, another case of a shift to NCA. The key to the success of this system lies in the variety of livestock feeding methods. In the past, possibilities to integrate crops and livestock were neglected and the contribution of livestock to the household income was limited. One way of improvement was to establish individual titles to land, visualised in demarcation and enclosure of grazing areas. Subsequently, some farmers developed the grazing areas to provide grazing, timber and fuel. They used multipurpose animals, they did not aim at fast maturity, they even accepted seasonal weight losses of their livestock (a typical approach within LEIA and NCA), but they aimed at high production on an area basis (an illustration of the concept of communal ideotype).

High stocking rates could be maintained through the use of crop residues, i.e. increasing population led to a reduced area of grazing land, a change in the role of cattle, and the replacement of livestock by specialised crops as the main source of cash. Adjustment was the key to sustainability, and population growth itself spontaneously responded to changed economic conditions between 1979 and 1989. Extreme shortages of land in parts of the district, combined with a national economic recession, high costs of education and other expenses for raising children, led to voluntary family planning. Finally, a programme in which people that had migrated to the city sent money back to the villages supported the process.

The evolution of the Machakos system differed between the drier and wetter areas of the district, and it depended on farm size:

— In the semi-arid region, animal draught power was important in various water conservation techniques, reducing risks of crop failure. Farmers on smaller farms combined wet season grazing with dry season use of crop residues, while larger farms relied more on grazing.

— In the subhumid region, where high population density leads to very small farms, draught power was no longer necessary and cattle were valued more for manure and milk.

In general, as land is scarce, fodder production was combined with soil conservation and stall feeding or tethering. Other methods of range improvement such as hedging, fencing, bush and indigenous tree management and scratch ploughing (superficial ploughing of the land), became attractive because they needed labour but almost no cash

Increased population led, via higher food demand, more labour, increased knowledge, and a reduction in the per capita cost of physical and social infrastructure, to an autonomous development towards higher agricultural production. The means and incentives to invest money and work in farm improvement and the knowledge of new and appropriate technologies were instrumental in this change. The new technologies came from various sources, such as traders, merchants, research and extension, religious groups, educated relatives and experimenting fellow-farmers.

Literacy and general knowledge were increasingly useful to find non-agricultural work, to make the most of a farming enterprise, and to participate in the various social and commercial networks. The development was assisted by government interventions and policies with respect to pricing, investments and education.

Government also provided community development services and changed research and extension from a top-down approach to one involving farmers in the development of technology. It supported small towns in creating jobs and infrastructure and becoming centres of trade and services for the rural area.

The Machakos case illustrates that farmers must be offered a variety of techniques from which they can select for managing pasture and other systems of animal feeding. The differentiation is necessary not only because agroclimatic conditions vary, but also because different combinations of grazing and stall-feeding may be economic for both small and large farmers.

Flemish/Norfolk System

This very intensive mode of integrated crop livestock farming originated in Flanders (Belgium) during the thirteenth and fourteenth centuries; it was elaborated in the United Kingdom in the seventeenth century. The system originates from conditions where the outfield/infield ratio declined due to increased population pressure. Land became scarce relative to the population and EXPAGR was replaced by NCA. Grazing of animals on the outfield to manure the infield became impossible and farmers were forced to find other ways of providing nutrients to their crops. They did so by keeping animals in a system of zero grazing on deep litter, quite similar to the Kano case. The animals were fed on crop residues and on crops that could fix or mobilise nutrients such as nitrogen and phosphorus.

Legumes fixed atmospheric nitrogen and rape crops (cruciferae) were used to mobilise soil phosphorus. The proximity of the city helped to generate cash from the sale of animal produce. Moreover, the use of night soil from the city on the crop fields was one additional way to maintain soil fertility by further recycling, i.e. integration and mixed farming at regional level.

Mediterranea Legume-grain Rotation

The effective use of legumes for nitrogen fixation in tropical areas is limited. The organic matter and associated protein quickly decomposes in conditions of high humidity and temperature, resulting in leaching and volatilisation of nitrogen compounds before the crop can use this valuable nutrient. Eventually this can even lead to acidification of the soils, an example of apparently sustainable systems that eventually outdo themselves. However, the Mediterranean climate has a peculiar characteristic that distinguishes it from hot and humid tropical ones.

The Mediterranean climate is characterised by wet, cool winters and hot but dry summers; it occurs in the United States and southern Australia, as well as around the Mediterranean. Under these conditions there is hardly any release of nitrogen in summer, and slow release of nitrogen in winter. In southern Australia this resulted in a sustainable crop rotation of grains and legumes, where animals feed on the biomass of legumes, straw and failed crops.

CIPAV system

A specific crop-livestock system based on a contribution of sugar cane/pigs and biogas has been developed in Colombia. It is called the CIPAV system, after the name of the institute where it was developed. It is being applied and tested in several parts of the world and the main ingredient is sugar cane, which produces the feed (juice and tops) and fuel (bagasse). Multipurpose trees and water plants such as duckweed supply the protein, while the trees also play other important roles such as controlling erosion, providing sinks for carbon dioxide and methane and as a source of biodiversity. Sugar cane and trees have well-developed systems of biological pest control; they can also do without much synthetic chemical input. Also they are easily separated into high and low-fibre fractions as required for the different end uses of feed for pigs, cattle and sheep and fuel.

The preferred animal species are pigs and ducks, which adapt readily to the high-moisture feed resources, and they have a high- meat:methane production ratio. Sheep, which can derive most of their feed from the cane tops and tree foliage, complement them. Buffaloes and/or triple purpose cattle can supply draught power as well as meat and milk. All the livestock are managed in partial or total confinement to minimise environmental damage and to maximise nutrient recycling to the crops. The CIPAV model is flexible, as witnessed by the increasing acceptance of many of the elements by resource-poor and entrepreneurial farmers.

2

Significance of Integrating Livestock with Crops

The world at large is not found not only on the natural environment but on managed lands such as farms and ranches. Farming can raise many environmental issues and can be considered the most important industry in the world. The pressure on farming can lead to many new ideas or ways to improve farming such as the integration of livestock. Livestock can be a great source for nitrogen, phosphorus, and potassium which may be used for crops or in a crop rotation. There are also many things to take into consideration when trying to integrate the two.

Animals have been essential in sustaining crop yields in many parts of the world where dung and draught from wasteland grazing are often a factor in the survival of a system. However, ways and objectives of keeping livestock are changing as a result of changing access to feed. Often, animal production can be associated with problems of unsustainability, which can be true in some cases, but in others it is definitely not.

Livestock is a major component of a system with long term sustainability. Animals have been essential in some cases for sustaining crop yields. Animals can help to sustain crop yields by increasing the rate of nutrient inputs and allows farmers to include crops that can fix atmospheric nitrogen, release immobilized phosphorus, or enhance organic matter. In many circumstances, applications of nitrogen, phosphorous, and potassium can and should be designed to provide for specific replacement level for minerals removed by the annual harvesting of crops. As Schiere stated in his studies, livestock can "further obtain food from cropping in the form of crop residues, failed crops, and fodder production from leys" where grazing by livestock usually takes place after deforestation and/or cropping. In all of these cases the key to the proper way to manage your soil depends on the proper amount of manure and soil tillage.

A key point when integrating livestock into a cropping system is trying to find the right mixture of the two. The three man categories are Low External Input Agriculture (LEIA), High External Input Agriculture (HEIA), and New Conservation Agriculture (NCA). The NCA is a system of farming that is a mix between LEIA and HEIA. NCA also uses livestock "to convert biomass from leys, bunds, alleys, and catch crops into economically valuable products". When converting form HEIA to NCA a major change is the reduction of fossil fuel. The aspects of crop and livestock integration depend on many circumstances.

Livestock has and can be an essential for the sustainability of existing farming systems. Livestock has been indispensable in sustaining crop yields in many parts of the world. The use of their dung and draught has been used for cultivation on the cropping field. The manure used as a compost creates high organic matter which is critical to many soil functions. Organic matter is also great for the soil because the higher the organic matter the higher the microorganisms and the higher the activity in the soil. A cow is one of the best composers out there because it is rich in nitrogen and even more so in the Northern climates. The mixed crop-livestock system is also a great way to recycle and use resources available to farmers. The fertilizers and fossil fuel are replaced by dung and draught which is cheaper and more economical.

Studies suggest that it is easier to maintain the yield level by reducing cropping intensity in mixed farming systems with livestock than in other farming systems compared to farming systems without livestock. Because of the prices and government subsidies to ecological farming, the economic results were equally good. Economically, integrated farming was less favorable than the other systems because ecological forage systems results in the least environmental harm, and is based on government subsidies.

The forage systems are also the most profitable because it gives the highest yields and best nutrition balances. *E. coli* is also a factor when considering using. manure because the feces growing in the manure may potentially contain E. coli. You also have to take into consideration that each farm is different and each system is different which forces the farmer to acclimatize to their situation.

Livestock can also play a negative roll in a mixed cropping system. The elements that are put in such as nitrogen and phosphorus can be hazardous if they are not effectively absorbed by the soil and utilized by growing vegetation. The run-off nitrogen and phosphorus from agricultural land will cause nitrate and phosphate contamination and toxicity in water. This is a problem that is so significant that the legislation is being enacted throughout Europe either to restrict inputs or nitrogen and phosphorus or govern the amount of nitrogen and phosphorus allowed on the land.

There are a few disadvantages and many benefits when it comes to integrating livestock into a cropping system. There is not only a general need for manures which can accommodate all the excreta that far animals produce, but also a specific need for animal excreta to sustain farming, especially organic farming.

Livestock's Contribution to the Environment

Over the last few decades, there has been much emphasis placed on the detrimental effects of livestock on the environment. Livestock have been blamed for water pollution in Europe, deforestation in South America, desertification in Africa and, at the global level, for increasing the greenhouse effect and for reducing biological diversity as well as for feeding on grain that could be better utilised directly by humans. These detrimental effects, however, are essentially generated by the way livestock have been managed recently, mostly on a short-term profit basis with no concern for sustainability.

In semi-arid and arid areas, previously existing management systems were disrupted by such factors as governmental ownership of rangelands replacing the traditional land tenure; uncontrolled practices of resettling crop farmers in areas only able to sustain livestock production in the long term; "plundering" of feed resources by outside speculators, often rich businesses interested in making high profits on a short-term basis; repeated droughts; common political will restraining the mobility of pastoral people; and also decreased mobility because of increased insecurity. All these factors interfered with the traditional balance that has permitted pastoral tribes to live for centuries within limited areas that were well managed.

Moreover, wherever it was possible the need for short-term profitability lead to the development of intensive animal production systems with no concern for the environmental impact. In these systems, sewage disposal was often a problem as it was too costly to transfer the sewage to crop areas. Consequently, a useful input for crop production was wasted and left to pollute water reservoirs. Like most other intensive mono-production agricultural systems, intensive livestock production cannot be sustainable and environmentally friendly if it is not integrated with other agricultural activities. It must not be forgotten that even though productivity and efficiency of livestock production is very low in developing countries, it can nevertheless be improved with considerably less inputs of fossil energy than is the case in developed countries.

Sustainable Livestock Production

Livestock's negative impact on arid and semi-arid areas has long been exaggerated and has only become controversial in recent years. Sidahmed and Yazman advocate that traditional pastoral systems are stable since they respond to high climatic variability. There are good prospects for animal production, which is fully compatible with environmental protection in these areas, as demonstrated by the following examples.

Recent studies on rangeland management in arid and semi-arid areas have shown that animal production can be perfectly sustainable if some precautions are taken. For example, to estimate the carrying capacity, a dry year should be taken as reference; for grazing capacity, only half of the biomass of herbaceous perennial; and for browsing, only 15 percent of the annual biomass production of palatable plants should be considered available. Other management aspects include fire control, surveillance of the presence of cattle around villages and watering points and protection of fragile soils during the rainy season.

In the case of Botswana and Zimbabwe, because the intrinsic resilience of rangeland was not acknowledged, much effort had been wasted in trying to stabilise production instead of promoting a "tracking" strategy that better followed variations in rainfall. Very often, traditional transhumant systems have proved to be much more sustainable and productive than ranch systems in the arid and semi-arid areas because of their high degree of flexibility. Nevertheless, as confirmed by Niamir, although a reversion to the traditional systems would be ideal, there is still much room for improving these traditional systems and developing locally appropriate techniques.

For dry lands, the traditional pastoral regimes are clearly superior to the mixed agricultural systems that agriculturists pressing northwards tend to replace them with, for example, in the Sahel. According to Perrier, rangeland management projects often fail because the wrong approach is taken. Modifications to local conditions, such as the replacement of traditional pastoralist models with centralised management, were usually undertaken in order to accommodate proposed innovations, mainly increased control of grazing resources. Such an approach often underestimated the complexity of the local production systems and usually only managed to disrupt these systems, generating conflicts and the deterioration of range management. Perrier proposed to revert this tendency by promoting range management proposals to fit the local context, such as repeat-seasonal grazing rather than rotational grazing, based on a systems approach, which would lead to a thorough understanding of the local production systems.

Thus, new prospects should arise for innovations in these systems, for example, the introduction of fodder trees and the use of supplements such as multinutrient blocks and urea-treated bush straw.

According to herders in the Niger, animals fed with such supplements are less selective when grazing on rangeland, thereby preventing the most palatable species from being overgrazed and, at the same time, contributing to the preservation of the environment. Another positive impact that ruminants have on vegetation is the dispersal and facilitated germination of fodder tree species. The growth of the seedlings is further assisted by the presence of nutrients in manure.

In arid and semi-arid areas, livestock are often considered to be only an animal production system, with no integration into other agricultural activities. But, in reality, the mobility of livestock in these systems usually implies great interaction with crop production systems, especially through manuring contracts between farmers and herders, providing draught animals and utilising crop residues as feed. Given the lack of chemical fertilisers in the Sahel, access to manure is of the utmost importance for promoting short cycle crop production and ensuring strengthened food security, despite rainy season uncertainties.

Integrated Livestock/Agriculture Production Systems

Most positive contributions of livestock to the environment are related to their role in integrated sustainable farming systems. Some examples are given below, although it is not intended that these constitute an exhaustive list covering all systems where livestock contribute positively to environment-friendly farming systems. Considered separately, a livestock production system generally demands less fossil energy than a crop production system does, which often relies heavily on fertilisers and herbicides that are manufactured and spread using expensive fossil fuel.

With the rapidly growing world population, the demand for grain is also increasing. Areas previously devoted to fodder crops have now been taken over by grain production, mainly to satisfy the needs of the population, with livestock relying more and more on crop residues and by-products, as is apparent in China and Egypt.

In Asia, through centuries of experience, the integration of livestock, fish and crops has proved to be a sustainable system. In China, for example, the integration of fishpond production with duck, goose, chicken, sheep, cattle or pig raising increased fish production by two to 3.9 times. Also recognised were the ecological and economic benefits of fish utilising animal wastes, a manageable system for small-scale farming.

Environmentally sound integration is ensured in these systems: livestock droppings and feed wastes can be poured directly into the pond, where they constitute feed for fish and zooplankton; livestock manure can also be used to fertilise grass, which also constitutes feed for fish; vegetables can be irrigated from the fish ponds, and their residues and by-products can be used for feeding livestock.

In Amazonian Ecuador, the introduction of sheep grazing on legume forage/fuelwood fallows within traditional farming systems has been shown to greatly benefit the environment. In this way, degraded lands can be rehabilitated for crop production and further use of less fertile lands can be reduced.

Maximising both integration and recycling ensures the sustainability of animal production/agricultural systems. These authors have shown that a sustainable livestock production system for the humid tropics can reach a high degree of integration. Such a system is now operational at the small-farmer level in Colombia and supports high levels of production. It includes pig, sheep, duck and earthworm raising (in partial or total confinement), food crops, environmentally protective perennial crops such as sugar cane, nitrogen-fixing trees (*Gliricidia septum* and *Erythrina fusca*) or other multipurpose trees (*Trichantera gigantea*) and water plants (*Azolla filiculoides*). Feed, energy (biogas and tree branches) and fertilisers (are all produced on the farm.

The different components of the role of livestock in a farming system are listed below.

Draught Animal Power

Livestock provide renewable energy for agriculture, saving a considerable amount of fossil energy that otherwise would be used mainly for manufacturing and operating heavy agricultural machinery, as well as for producing fertilisers. For example, it has been calculated that the energy needed for soybean seed-bed preparation and sowing using a tractor is 6.3 times more than that required when using draught animal power. Moreover, the latter is renewable, while this is mostly not the case with the former, and this figure does not take into account the fossil energy spent making the tractor and maintaining it. To a large extent, draught animal power is self-repairing, self-propagating and sustained by by-products produced on the farm. Draught animal power can also play a key role in large-scale production systems, as is the case on a 30,000 ha sugar estate in the Dominican Republic, where some 18 000 oxen haul sugar cane from the fields to the railway system leading to the sugar mill. Sustainability is guaranteed in such a system, since the oxen's main source of energy comes from the sugar-cane tops and, in the end, the manure goes back to the sugar cane.

In many parts of the world, draught animal power is often the best alternative leading to increased yields of crop production. As pointed out by Delgado, the expansion of cultivated area is often limited by a shortage of labour for weeding and harvesting needed for a larger crop output. In the long term, more extensive use of draught animal power for weeding and harvesting will lead automatically to the expansion of cultivated area. Compared with human traction, this type of system contributes to the improved quality and timely execution of farming operations, such as tillage. It also permits the introduction of soil conservation practices to impede excessive soil compaction and erosion.

Fertiliser Production

In mixed-farming systems, livestock are often fed on crop residues and by-products as well as pasture. In turn, some nutrients and organic matter are returned to the soil through livestock manure ensuring the maintenance of soil fertility and acting as a soil conditioner. The main results are improved cation exchange, better absorption of water and the prevention of runoff and soil surface crusting. On the other hand, the exclusive use of mineral fertilisers will result in a decrease in base saturation and pH and in the occurrence of aluminium toxicity. It is therefore clear that manure plays a key role in sustainable crop-livestock production systems. Moreover, small farmers in developing countries often cannot afford any other type of fertiliser, and fallow periods are becoming shorter as population density increases. The direct result of the proper use of manure is increased yields, which may prevent small farmers from further encroaching on fragile areas.

The value of manure as fertiliser could be increased through better animal feeding. There are now various widely spread techniques to upgrade roughage, such as urea treatment, which was practiced in China in 1993 by 3.3 million small farmers, who treated more than 11 million tonnes of straw. It was observed in the Niger that the percentage of crude protein in manure increased from between 4.5 and 5.8 percent with a traditional diet to as much as 7 to 8.3 percent with urea-treated roughage, which allowed the farmers to decrease the use of costly, often imported chemical fertilisers. Also in the Niger, farmers using the upgraded manure could reduce the urea used for rice by half and still increase their yields by 40 to 50 percent.

Weed Control

Livestock can be successfully used for weed control and therefore contribute to the decrease of water pollution by herbicides. The fossil energy that would otherwise be used for making and spreading herbicides is also saved. Livestock are currently used in sugar-cane, oil-palm, coconut, rubber, citrus, kola and mango plantations. While grazing in these plantations, the livestock, most commonly sheep, also provide organic fertiliser. On a100-ha sugar-cane estate in Colombia, for example, a flock of 175 African hair ewes control the weeds, replacing the need for either tractors or herbicides. Other species also useful for weed control are geese, which are used in India, and horses, which, because of a herbicide shortage, have been used successfully in Cuba for the last three years in citrus plantations to control such weeds as *Panicum maximum* and *Hyparrhenia rufa* that have a very negative effect on orange yields. Fish, such as grass carp and silver carp, are very efficient for weed control in drains and canals.

Table 1. Relative dung and nitrogen production for different animal species (per head per day)

Species	*Adult live weight (kg)*	*Dung production (kg dry matter)*	*Nitrogen content (%)*	*Nitrogen production (g)*	*Nitrogen production (kg/year)*
Buffalo	460	5.8	0.80	46.4	16.9
Cattle	350	4.4	0.73	32.1	11.7
Goat	20	0.3	1.32	4.0	1.5
Sheep	20	0.3	0.91	2.7	1.0
Chicken	2	0.05	3.9	2.0×10^{-4}	0.07
Duck	3	0.06	3.0	0.18×10^{-4}	0.07

Livestock production is integrated with forestry mainly for reforestation purposes or to prevent forest and bushfires. In Scandinavia, sheep grazing is used in reforestation areas to control shrubs that tend to suffocate young trees. Goats have been used to create fire breaks and are also widely used to control weeds. In Europe, the European Union provides some countries with subsidies to encourage grazing in the woody areas of the Mediterranean in order to prevent fires.

Source of Fuel

The use of cow dung as fuel in India and other countries results in an enormous amount of trees being saved, and its ash can be used as fertiliser. When water is not a limiting factor, however, dung can be utilised much more efficiently for fuel production through biodigesters.

Zero-grazing livestock production systems can be environment-friendly when feeds are grown on-farm and slurry is collected in biodigesters, the effluent of which may be used either as crop fertiliser or for fish production in ponds. For the former there is the added advantage that weed seeds have been inactivated during the digestion process.

The Grameen Bank in Bangladesh - which is very famous for its successful credit schemes for the rural poor, especially the "cattle lease" programme for many women - is now linking this activity to a biogas programme since the biodigester is considered to be the centrepiece of an integrated farming system, where the effluent entering ponds increases fish yields by accelerating algae production. Since very cheap and simple biodigesters made of plastic sheets have recently been developed through FAO projects, they are now more accessible for rural families than the more expensive cement or steel ones.

In Cambodia, an FAO project has recently demonstrated that, where there is adequate manure from at least three cows or eight pigs, the amount of gas produced is sufficient to replace about 75 percent of the fuelwood normally used by an average family of six persons. On-farm production of biogas, therefore, can satisfy most of the basic energy needs of a household, at least for cooking. This would obviate much of the need for fuelwood, which is becoming a scarce resource in many developing countries and must be gathered very far away, involving high energy costs for collection and transport, while jeopardising the environment through excessive collection.

Role of Feed Base in Environmental Protection

The feed base of animal production can play an important role in environmental protection. It consists, of either the main energy sources, such as food-crop by-products and residues and sugar cane, or the main protein sources, such as multipurpose trees and water plants.

If not used as animal feed, agricultural residues such as straw, stover and groundnut or cowpea hay are burned, polluting the atmosphere. Some agro-industrial by-products are already widely known as feed, for example, cereal brans, oilcakes, molasses, brewer's yeast and beet pulp. Others, such as sisal leaf waste, filter-press mud from the sugar industry, coffee, citrus, pineapple or tomato pulp, olive cakes and palm oil sludge, are usually produced in large quantities as well, and their disposal is often a problem. Most of these by-products could be used as feeds.

By-products from livestock industries such as slaughterhouse wastes (poultry offal, bones, blood), whey from the cheese industry, hatchery by-products and fish and shrimp wastes can be utilised in ruminant feeding. Even excrete from one animal species can be used as feed for another. Poultry manure or litter, which is rich in crude protein, can be fed as silage to ruminants after processing to ensure its sterilisation.

In Cuba, domestic organic wastage derived from human consumption and other wastes originating from the food industry are collected throughout the country and cooked on an industrialised basis in small factories to feed pigs reared in the vicinity.

High biomass and perennial fodders (sugar cane and trees) act as carbon sinks, counteracting the greenhouse effect. They also impede erosion. The best use of these fodders permits the reduction of land area per livestock unit.

Legumes, such as fodder trees, planted together with crops or used as ground cover in coconut, rubber or oil palm plantations, play a key role by fixing atmospheric nitrogen and increasing water absorption in the soil, which prevents erosion and decreases the need for artificial fertilisers.

Multipurpose trees integrated into a livestock production system can contribute to the protection of the environment in various ways. They impede erosion and increase nitrogen in the soil (nitrogen-fixing trees), thereby decreasing the need for agrochemicals (fossil-energy consuming). They can also supply domestic energy, such as wood, together with biogas, and preserve local vegetal biological diversity. There is already a large list of useful multipurpose trees known throughout the world, such as *Gliricidia* sp., *Prosopis* sp. and *Leucaena* sp. Nevertheless, there are many other species that would be worth propagating, although these are still to be identified through better investigation of indigenous knowledge.

Water plants, such as *Azolla* sp. and *Lemna* sp., also play a key role in animal production systems. They are a protein-rich feed, but they are also useful for purifying surface waters polluted with organic matters (livestock and human wastes), as carbon sinks and for fixing atmospheric nitrogen (azolla).

All these components also contribute to increased biological diversity within farming systems, as compared with modern intensive agricultural production systems, especially through the reduced need for pesticides. Even the environment-friendly alley farming system (nitrogen-fixing trees and cash crops) gains in sustainability if associated with animal production through the use of foliage as feed. This is confirmed by the recently developed three-strata forage system in the dry areas of Bali in Indonesia. This system includes, on the periphery of small square crop fields of 0.16 ha of maize, soybean and cassava, the use of three species each of grasses and ground legumes (as first stratum, 0.09 ha), and a 200-m circumference area with three species of shrub legume (as second stratum, 2 000 plants) and three species of fodder trees (as third stratum, 42 plants) in order to supply livestock with green feed throughout the year.

The advantages of this system are increased livestock production; increased biological diversity; production of fuelwood at farm level; greater returns from organic matter and nitrogen to the soil and prevention of erosion caused by excessive grazing of crop fields after harvest; and better results with scavenging poultry, honey and snail production. Taken together, these contribute to the protection of the environment and the sustainability of the system.

Other Positive Impacts of Livestock on the Environment

There are fragile areas, in the north of the Sahel, for example, where the integration of livestock into farming systems, by ensuring much more regular income than from crop production alone, can provide small farmers with enough security so that they are not forced to expand cultivation into areas where the probability of harvesting is very low but the erosion effect is high.

Most crop production systems generate seasonal incomes, but the integration of livestock production would result in supplementary and regular incomes from the sale of milk and eggs, as well as provide emergency incomes, whenever necessary, from the sale of animals.

This additional income would also allow farmers to make their production systems more sustainable and environment-friendly through, for example, conversion from the slash-and-burn system to the sedentary production system, better preparation and fertilisation of crop fields, implementation of soil protection measures such as terracing to avoid erosion, the use of selected seeds and planting of legume fodder trees These investments could not be afforded otherwise. The integration of livestock into crop production systems also offers increased prospects for local labour, which would slow down the drift from the land and the never-ending expansion of big cities.

3

Livestock-Fish Farming Systems

Understanding how farming households meet their needs is essential to assess the likely adoption of fish culture. A study of the evolutionary development of farming systems provides a useful framework since the nature and intensity of farming activities may indicate the likelihood of fish culture being appropriate. We analyse the factors that have stimulated intensification of farming and relate this to prospects for integration of livestock and fish production. Agricultural development has been linked to the pressures of human population growth and we also examine this effect, together with the impacts of changes caused by urbanisation and industrialisation.

Settled agriculture is divided into three phases to indicate the potential role of integrated farming, particularly with respect to smallholder farmers in less developed countries (LDCs). Pastoral nomadism and shifting cultivation have limited aquaculture potential. Pastoral systems occur in arid regions in which seasonal availability of grazing limits carrying capacity of livestock and nomadism is a necessary part of a livelihood strategy. This movement of both pastoralists and shifting cultivators has necessarily constrained development of fish culture.

The limited extent and duration of surface waters in arid regions also constrains natural fish production; and fish consumption is usually unimportant or absent among peoples living in such areas. Harvesting wild stocks generally remains important for aquatic foods after settled agriculture is well developed, whereas hunting and gathering terrestrial food declines rapidly as agriculture evolves.

The evolutionary development of both livestock and fish production can be classified within a schema derived from the same broader farming systems context. Settled agriculture is typical of many pre-industrial societies where there is little integration between crops and livestock managed principally for draught and to meet social obligations. If fish culture occurs it is normally at a very extensive level and closely associated with management of wild fish stocks. In settled agriculture II, the production of livestock and crops are more intimately linked, with livestock fed on crops and crop by-products and their manure essential for maintaining soil fertility.

Use of N fixing plants together with other inputs such as nightsoil is also a common strategy for maintaining productivity. It is in this context that most traditional integrated fish culture is found. The trend towards industrial monoculture is a model followed by both livestock and fish production and is widely adopted in developed economies. Recently however, environmental concerns about heavy use of agrochemicals and waste disposal, consumer pressure and legislation are leading to some return to a more balanced approach to food production.

The tendency to develop more intensive farming systems that produce more food per unit area per unit time has traditionally been linked to increasing population pressure. Global population is expected to rise further from the current level of 6 billion, and may double before stabilising, even at the most optimistic projections. Global population is split more or less equally between urban and rural areas but urban areas are expected to surpass rural area in population around the year 2005 and to account for 60 percent of the total by 2020.

Population pressure has not occurred, nor exerted its impact on intensification of food production evenly. Historically more fertile, well-watered environments have had greater productive potential and supported higher human populations. Thus, well-endowed floodplain agroecosystems in Asia have become the site of the most intensive traditional agricultural practices.

Globalisation of trade pre-dating the colonial era, industrialisation and major changes in human medicine have fundamentally de-linked food production and human population densities. If the concept of agro-climatic population, or the population in terms of food production capacity is used, today semi-arid zones are typically under much greater population pressure relative to land endowments than humid areas. Although historically most of Africa has had little pressure on land resources, by 2025 the majority of the continent will comprise high-density countries requiring highly productive agricultural techniques.

Accelerating urbanisation has stimulated demand for industrial food production in both developing and developed countries. Industrial food production requires intensive applications of resources, particularly energy, nutrients and water and is dependent on scientific knowledge.

Farming operations spanning a wide range of intensity levels can be found increasingly within the same country although some doubt that coexistence of less intensive production systems with industrial methods is possible over the long term. Integration of livestock and fish, in which one or both subsystems does not become entirely agro-industrially based, may better fit the limited resource base of smallholders and improve environmental sustainability.

Environmental Effects

Environments have shaped cultures and dietary norms and taboos that in turn explains current distribution and dependence on livestock and fish. Recent anthropological research has shown that the concepts of the 'sacred' cow and 'abominable' pig have an environmental basis. The advantages of ruminants that digest cellulose and thus do not compete for food with humans, together with their more multipurpose attributes, are the bases for the cultural bias.

The rejection of fish as food is also common among people in arid environments where surface water and natural stocks of aquatic animals are rare. The lack of large livestock in traditional slash and burn-based societies in Africa, and elsewhere, can be related to a low requirement for tillage, and their poor survival because of the tsetse fly. Animal protein needs could be met by the harvest of game and wild fish and intensification of livestock and fish production was unnecessary at the low human population densities found in typical swidden agricultural societies.

If natural supplies of wild stocks are particularly rich, much higher population densities may be supported, provided human dietary energy needs are met. The rice-fish societies of lowland Asia are good examples of this situation where diets based on cultured, calorie-rich rice were balanced by diverse, aquatic plant and animal food gathered from the floodplains. Indeed, whilst natural fish supplies remain adequate, there is little interest in fish culture. Seasonal inundation of flood plains that led to dependence on aquatic-based food sources probably also limited the importance of livestock because of seasonal shortages of feed.

In contrast, arid environments have stimulated pastoral systems in which low densities of ruminant livestock are grazed on extensive, common property pastures. The challenges associated with increasing productivity in such marginal, community-based resources systems are similar in both water-short, livestock-based pastures and water-rich, communally exploited wetlands. Major issues include how intensification can occur whilst safeguarding equity and the environment.

Crop Domination

In contrast to most shifting farming, in which livestock is relatively unimportant, or pastoralists in which livestock dominate, animals fulfil small but important roles within the household in settled agriculture. Furthermore, settled agriculture has much greater potential for aquaculture. Most land is reserved for crops and livestock are kept mainly for draught in settled agriculture phase I. Pigs and poultry may also be kept in small numbers, and usually scavenge and are fed wastes from the household. There is little integration between crops and livestock, largely because the number and nutritional status of the livestock are low.

Ruminants depend mainly on limited rough grazing of harvested fields and common land. Limited crop diversity, as well as little recycling of crop residues and manures, are characteristics of crop-dominated systems. Such resource-poverty is typical of most small-scale farmers in developing countries today, except those that have leapfrogged to settled agriculture III through the green revolution.

Crop-dominated systems include crops grown as the dietary staple such as rice and maize, often for subsistence, together with other crops grown for cash. Various levels of intensification may be evident e.g. irrigation, terracing, fertilisation and weeding. Orchard crops and vegetables are often grown in home gardens.

In some parts of the developing world, such as Southern Viet Nam, both livestock and fish are relatively important even in crop-dominated livelihoods but the potential is far. from realised by most households which rely mainly on wild fish. One survey indicated that farms in Central Thailand, which had diversified away from rice monoculture were more likely to use animal waste for fish culture than farms continuing to concentrate on rice production.

Such intensification of livestock in settled agricultural phase I is often limited by poor feed availability. Also, draught animals tend to be used irregularly so although their productivity is low, farmers have little interest in improving their performance. Parallel development of intensive feedlot operations may also reduce opportunities for small-scale pig and poultry production.

Widespread adoption of fish culture may not have occurred within crop-dominated systems, even when fish are valued and consumed. Stocks of wild fish may remain at a level that satisfy rural peoples' needs. Poorly developed on-farm water storage, or a lack of seed or knowledge may also constrain adoption.

Traditional aquaculture in the valleys of upland areas of Indochina and Southern China, where population densities are high and wild fish stocks are very limited, suggest that aquaculture can evolve under these conditions, but that linkages between livestock and fish were relatively weak. Intensification of fish production based on animal manures would be constrained by the limited numbers of livestock and difficulties in collection and use of their waste. Livestock diseases also constrain inventories of livestock in many instances.

Disease constrains livestock productionAquaculture is a recent development among certain ethnic minorities, such as the Hmong in upland areas of Indochina, for whom pig production is both traditional and important. Under current conditions disease is a greater constraint to expansion of pig production than feed availability; more arrowroot or cassava can be grown, or vegetables and banana stems cut from the forest, if supplies of maize are limited.

Pig wastes are used extensively, with wastewater directed towards opium-poppy growing plots, but the siting of pens over fish ponds has been adopted by some households. Small flocks of mixed poultry, allowed to scavenge for natural feed and given small amounts of supplementary feed such as paddy grain or ricebran, are common in Asia.

Integrated Crop/Livestock

The integration of livestock with crops, or mixed farming, is the major characteristic of settled agriculture phase II. Livestock fed arable crops and improved pasture produced on the farm is the main focus. Crops are intimately integrated with livestock as manures are used to maintain soil fertility together with N fixing legumes.

Much of the farming in Western Europe and Eastern USA was of this type between 1850-1945. However, the recent increased control of nutrient effluents has begun to favour this form of farming again over industrial monoculture.

The origins of mixed farming lie in increased demand for livestock products from urban centres. Various methods were adopted to increase livestock such as production and feeding of turnips to livestock during the winter and the rotation of cereal crops with legumes such as clover. More inputs such as nightsoil from urban centres, followed by inorganic fertilisers and feed concentrates, increased livestock densities and soil fertility.

Integration of fish culture into farming systems has developed in areas where ponds were essential to diversification of rice-dominant systems and livestock were relatively few and feed limited. This has occurred in flood-prone areas, often where rice yields were low and land was raised to make dikes for planting perennial or upland crops.

Increasingly, buildings and roads are constructed on raised dikes and fill is obtained from borrow pits. The ponds excavated often serve primarily for storing water on-farm. Expansion of on-farm reservoirs (OFRs) has also expanded in areas in which drought otherwise constrained any intensification of cropping.

Analysis of traditional integration of fish production within the highly diversified farms in the Zhujiang Delta, Southern China, indicates that wastes from livestock (pigs, silkworms) and people were important inputs. Fish production, however, was mainly based on the feeding of wild, uncultivated grasses for the macrophagous grass carp. This fish species largely filled the niche occupied by ruminants in mixed farming systems in Europe.

Recently other macrophagous fish species such as the silver barb have been promoted to utilise seasonally abundant duckweed in Bangladesh. The feeding of leafy vegetable material is traditional for raising macrophagous giant gourami in Indonesia, and potential exists for similar systems elsewhere based on herbivorous *tilapias* and *Colossoma* spp.

The major constraint to fish culture within farming systems based on leafy vegetation is the availability of adequate amounts to meet the needs of growing fish. Constraints to, and opportunities for, intensification of macrophagous fish production are similar to ruminants. Opportunistic use of crop harvest by-products would not normally provide consistent levels of feed or feed quality.

Continuous cropping of arable crops, e.g. cassava leaves has trade-offs in terms of the main crop yield. The use of arable weeds from intensive horticulture has unrealised potential in some situations but would normally be constrained by irregularity of supply. Where such products are available, there will usually be competition with livestock.

Industrial Monoculture

Many 'modern' settled agricultural farms stages I and II have intensified production by adopting some aspects of the scientific-industrial 'revolution'. Industrial monoculture has evolved to supply ever larger, more concentrated markets with homogenous products. Increasingly dependent on the fruits of science and engineering, traditional mixed farming has changed over the last fifty years, using a greater range and volume of inputs. Improved varieties, agricultural chemicals, feeds and mechanisation are used to produce fewer products in greater volume; many farm operations have become monocultures.

The technical complexity and economies of scale characteristic of industrial agriculture encourage this tendency. In most cases the ready availability and low cost of industrial nutrients has reduced the need for integrating crop and livestock production. Industrial aquaculture, often of carnivorous species, evolved from using local surpluses of trash fish of little value to fatten wild fish and there were few links with land-based agriculture.

Two important reasons suggest that industrial monoculture will evolve towards greater integration with other food production. Firstly, the real costs of adverse environmental impacts of specialised production are now becoming clear and closer integration can reduce or eliminate them. Secondly, wastes from intensive animal production can be valuable inputs into other parts of the farming system. Apart from maintaining soil structure in arable systems, these include their direct and indirect use in fish production.

In parts of Asia where concepts of waste recycling are traditional and well understood, the industrialisation of agriculture and changing demand are both challenging and opening new possibilities for integration. On the one hand integrated livestock-fish systems are evolving in China to rely increasingly on wastes from nontraditional livestock such as dairy cows and broiler chickens. Livestock wastes are also being used for a greater range of purposes; mushroom, maggot or earthworm production may be more profitable than aquaculture. However, rapid industrialisation and moves towards intensive aquaculture practices can also undermine traditional integrated practices and threaten ecological stability.

Types of Integrated Systems

The role of livestock in the intensive, rice-based farming systems on the major floodplains of China, where aquaculture first developed, was fairly minor and linkages between fish and livestock production correspondingly weak. Applying the principles of waste recycling was, however, integral to sustaining China's high population densities. In recent decades, as the importance and concentration of livestock and their wastes have increased, these ideas have been applied more widely in China and elsewhere in Asia to reuse livestock waste for fish production. A range of experience and systems now exist.

The growth in production of monogastric livestock and agribusiness concerns has had a major impact on availability of waste for aquaculture. It has also changed the manner in which livestock are raised and marketed. Lower priced livestock products are the result but most are produced by entrepreneurs. Small-holder farmers have often benefited least.

Attempts to scale-down commercial livestock-fish systems for resource-poorer farmers have been uneven because of difficulties of access to, and management of, inputs such as feed and outputs. Alternative livestock systems that produce waste suitable for fish culture are still largely undeveloped. Typical smallholder, ruminant and monogastric production is extensive, making wastes difficult to collect and use. The wastes from poorly fed animals are often low quality. Upgrading such livestock systems can increase the potential for integration with fish culture. Strategic intensification and integration of livestock and fish are alternative paradigms to the conventional models of agribusiness development. Improved efficiency of nutrient use under conditions of nutrient scarcity can be a major benefit of integration.

Least-cost production techniques that depend on using several different fertilisers and feeds, rather than a single high-cost compounded feed, is a relevant strategy for both livestock and fish production. Commercial livestock production and processing typically produce nutrients in abundance, requiring cost-efficient disposal. This offers a different but equally important rationale for integration.

Livestock-fish systems in which waste from animals continuously raised in a pen and fed complete feeds ('feedlot') are used as pond fertilisers are the most common type of integrated system outside China. Important characteristics used to classify such systems include type of livestock and fish species and scale of operation. The degree of integration has already been raised in definitions of integrated farming. Definitions can include whether the fish sub-system receives all the livestock wastes, if livestock waste is used as a single input or as part of a multiple-input system. Since fish production is closely linked to the quality and quantity of inputs, the nutritional value of specific livestock wastes, in fresh or processed form, is also an important descriptor.

The major types of livestock may be classified on their potential for integration with fish culture. Most fish production data for culture systems fertilised with livestock manure are based on feedlot systems but wastes from animals raised in scavenging systems can also be used. Clearly the latter waste is less easy to collect than waste from animals confined continuously. Some livestock wastes are more acceptable than others; whereas poultry manure from broiler houses is easy to handle and is used widely in agriculture, pig manure is often less acceptable for logistic, religious or aesthetic reasons. This is also reflected in opportunity costs. These may be relatively high for dried poultry waste and low for pig and cattle slurry. Nutrient density is also an important factor as monogastric animals fed high-quality feeds tend to have manure that is more nutrient dense and valuable than ruminants, particularly those raised on nutrient-poor tropical volunteer vegetation.

Some animal manures contain compounds that cause water quality problems, such as tannins and indigestible fractions. Waste from animals receiving poor nutrition, especially those on diets with high levels of fibre, is particularly unsuitable for fish culture. Other wastes may be contaminated with deleterious compounds such as pesticides, drugs or feed residues that have negative impacts on quality.

Monogastrics

Most integrated livestock-fish systems described in the literature and operating successfully involve the use of manure from feedlot monogastrics. Pig and poultry manure is used fresh, or commonly for poultry, in a dried form for semi-intensive fish culture. Design criteria are given below.

Pigs

Native varieties of pigs are well adapted to existing on roots, leafy vegetables, cereal by-products and household food waste. Although they are not ruminants, fermentation in the hind gut is significant and this allows them to grow slowly on a poor diet. An important food item in some cultures is human waste, scavenging pigs playing an important role in nightsoil disposal.

In some cases intensification has been built around waste recycling. Traditional pig production in China was mainly for the production of manure as is still the case in northern Vietnam. Weaned piglets are traditionally fattened on considerable quantities of grain byproducts or root crops, especially sweet potato supplemented with cultivated and volunteer green fodder.

The number of animals raised is therefore related to the production and availability of these feeds, and the number raised per household is low. Pig manure was used to fertilise rice and vegetables, which may explain why the average fish yields in China were as low as only 2 tonnes. ha-1 until as recently as two decades ago, before more widespread introduction of feedlot livestock-fish integration. Fish culture in northern Vietnam is similarly constrained by the competing use of pig manure to fertilise crops.

Poultry

Local breeds of poultry can thrive in environments with abundant natural feeds, which are often rich in calcium and animal protein. Ducks are particularly important in the flooded agroecosystems of South Kalimantan and Southern Vietnam, and in Java where irrigation supports year-round, paddy rice production. Management has become more intensive in many cases. It involves the confinement of poultry for all or prolonged periods to ensure both adequate nutrition and production, and control damage that freely scavenging animals can inflict on crops.

Poultry are also raised in large numbers when local or seasonal food surpluses occur. An abundance of aquatic snails from the lake allowed duck producers to intensify balut duck egg production in Laguna, Philippines, before feed mill-based feeds were available. Meat ducks are purchased and raised for short periods in large quantities to grow on dropped grain in ricefields after the rice harvest in Central Thailand during which time there is little opportunity for waste collection. But prior to slaughter the ducks are fattened intensively at high density producing large amounts of high quality, easily collected waste that is used for fish culture.

Small-scale, feedlot monogastric-fish systems have been shown to be technically viable in on-station and researcher-managed, on-farm trials. However sustained adoption of these types of systems by farmers has been rare. Almost all research has been carried out using waste from 'improved' livestock systems, typically feedlot production of pigs and poultry. Such research has normally been carried out without participation of farmers in the design stage and, often, with financial incentives to encourage collaboration.

The normal criteria used for assessment of 'viability' are positive internal rates of return, or return to labour or capital. However, these are not holistic enough indicators, given the complexity of small-scale farmers' livelihoods. Resource-poor farmers have great difficulty sustaining even scaled-down, feedlot integrated systems. In Thailand it was found that production costs, particularly feed costs were high and there were problems in both securing inputs and marketing produce in rural areas. In effect, integration of feedlot livestock and fish for most small-scale farmers is a 'technological mismatch'.

Small is a relative term for both fish and livestock production. A profitable layer chicken-fish system was described in one area of Northeast Thailand. Although farm size was within the norm for the region, operations tended to be irrigated and highly capitalised compared with typical, rain-fed farms in the same area. The concentration of operations close to a provincial urban centre also favoured the provision of cost-effective inputs and sale of produce, and the scale of operations was advantageous to both farmers and middlemen.

The introduction of new types of livestock without consideration of markets has been another cause of failure in livestock-fish systems. Cattle and broiler chicken production were 'standalone' enterprises in Panama before fishponds were introduced, whereas pigs and ducks were introduced with the fish. The latter subsequently suffered marketing problems because of low demand, which consequently had an inverse impact on fish yields as there were insufficient livestock to fertilise the ponds. Similar problems have been reported in various African countries where meat ducks were introduced specifically to provide manure to fertilise fishponds without due attention to the wider system.

Ruminants

Ruminant faecal waste is probably the most commonly used fish pond input in developing countries. In addition to its use in fish culture being limited by competition as a crop fertiliser or fuel, it has an inherent poor quality as a pond input. Wastes from unimproved grasing animals may be too poor in quality to produce acceptable fish yields in small ponds. In an on-farm trial, around 4 tonnes of fresh ruminant manure had to be collected and used to produce only about 20 kg of fish in a backyard 200 m^2 pond. The buffalo manure had low N and P contents and also contained tannins which stained pond water brown, further limiting phytoplankton growth due to reduced light penetration into the water column. Tannins are also known to be toxic to tilapias.

A further major constraint is the collection and transport of low-value, bulky ruminant manures. Fish culture, however, may be integrated with traditional livestock management practices through linking the needs of livestock for watering, wallowing or overnight corralling. The need to handle and move bulky ruminant manure may thus be negated by strategic husbandry practices. Communal or individually-owned water bodies designed to allow the walk-in entry of large ruminants without physical destruction of embankments can allow low-cost ruminant waste input as well as conservation and concentration of nutrients from the watershed as a whole.

Temporary overnight corralling and/or midday wallowing for water buffaloes in small areas may further concentrate wastes. Farmers can therefore 'harvest' nutrients from their own and others livestock at little cost. The relationship between herders and crop growers in West Africa in which livestock are rented to deposit dung overnight in an area to be planted to crops gives a working example how manure could be managed in fish culture. In Northwest Bangladesh, the overnight tethering and supplementary feeding of cattle around the house allows drainage of urine into nearby borrow pits, enhancing the productivity of an important source of fish consumed by the household.

Non-conventional livestock

Non-conventional livestock encompasses both 'micro-livestock', a term coined for species that are inherently small and seldom considered in the broader picture of livestock development and invertebrates, which may be produced as either direct food items for humans or constitute part of the diet of other livestock or fish. Over 40 multipurpose species with promise for smallholders were identified in an exercise involving 300 animal scientists in 80 countries. Many will produce wastes that can be readily collected because of their small size and confinement during production and are suitable for aquaculture.

Small individual size, and limited demands for food and space make micro-livestock increasingly attractive to resource-poor people. Micro-livestock, including both rare breeds of domesticated species and wild animals with potential for domestication, often retain local importance to small-holder families and have potential for larger markets, especially in cultures that value wild or bush meat.

Non-conventional livestock that can grow and reproduce on feeds that do not compete with human or other livestock are particularly interesting. Some invertebrates can be included in this category and are locally important as sources of food and fibre.

Pasture/crop by-product fed

These include geese and rabbits for direct human consumption; and silkworms which are mainly raised to produce high value, natural silk fibre. Geese and rabbits have played important roles in small-holder systems in which pasture and field crop by-products are available. Silkworms have been an important component of small-scale farms in parts of Asia for millennia.

The high intake, poor digestion strategy of geese means that their manures would probably function in a similar way to that of the grass carp in fishponds. In contrast, the copraphagy of the rabbit may reduce nutrient levels of their manure; rabbit manure, however, is widely prized as a fertiliser and has been found to be a useful fish pond input in Africa.

Woody waste

Termites are some of the most commonly used feeds in recently developed smallholders fishponds in parts of Southeast Asia. However, unsustainable harvesting of wild mounds normally quickly eliminates this potentially important food source as its rate of regeneration is usually far too low to permit even medium term use.

However in Ghana, harvesting of termites for poultry is carried out sustainably by placing moist cow dung over termite nests shielded with tins. The termites burrow into the dung and some can then be fed to the chickens daily. Although actual quantities are small, termites are high quality feeds for a range of fish species and poultry. Where woody waste is produced in large amounts, their controlled production is a possibility.

Upgrading of Livestock Systems

Changes in traditional livestock systems have direct implications for peoples' welfare, biodiversity and the wider environment. The place of livestock within farming systems has always been dynamic and, particularly in marginal environments, reflected the resource base. Intensification of animal production has often been a critical part of improving rural livelihoods generally and a key issue is whether upgrading of livestock can become a key part in promotion of rural aquaculture. 'Upgrading' suggests an increase in productivity and importance of the animal within the farmer's system; normally genetic and management changes may constitute upgrading. Changes in germplasm alone rarely support long-term change.

The upgrading of livestock systems in developing countries has often resulted in a total loss of traditional breeds and management methods. A focus on improving yields rather than meeting the needs of the farmers and consumers has often been counterproductive in Least Developed Countries. Typically efforts have been made to develop a modern poultry industry in parts of Africa, whilst smallholder production has been largely ignored. However, improved management, disease control and strategic supplementary feeding can vastly improve productivity.

Sometimes upgrading has occurred in response to markets or demand changes; the poorer meat quality and body shape/size hastened the decline of the indigenous Mongkon pig in Viet Nam despite their advantages in terms of tolerance to heat and low input diets. Farmers also prefer modern breeds because they grow faster whether fed traditionally, as in northern Viet Nam, or mainly on formulated feed in the south of the country. There has often been a failure to fully understand the complex role of traditional livestock production in rural livelihoods. After the purposeful destruction of the native creole pigs in Haiti because of fears of their possible role in transferring African swine fever to North America, modern breeds of pig failed to meet small-scale farmers' needs and a synthetic version of the traditional breed had to be 'reinvented'.

Livestock systems, however, generally have to be improved in terms of the collection and quality of waste for productive integration with aquaculture. In practice, livestock manure from unimproved systems may be too small in quantity, too poor in quality, or simply unavailable because of competition from alternative uses, to be of relevance for aquaculture. Where livestock/crop interactions are already under threat, for example in Nepal, sustainability of crop yields are already dropping because of intensified practices and declines in livestock. In contrast, increased demand for milk and intensified cropping of a fodder crop rotated with cereals is leading to more livestock and greater sustainability in the Punjab.

The quality and quantity of wastes are influenced by the diet and other factors. The collectability of livestock wastes depends on feeding and housing arrangements. Clearly, waste can be collected more fully and efficiently in animals confined for all or most of the production cycle. However, larger numbers of animals raised more extensively, but confined for part of the production cycle, can also produce substantial quantities of waste. Moreover if quantity and quality of feed consumed during the period of grazing/scavenging are high, this is reflected in the quality of the waste.

Several factors mitigate against the ability of smallholders to increase number, confine and feed livestock enough to benefit from using their waste for fish culture. The limited ability of monogastrics to utilise diets high in fibre has reduced the scope of improvements to traditional systems of raising pigs and poultry. The number of monogastrics that could be raised has been linked to the availability of grains and/or root crops surplus to direct human consumption needs.

The processing of staple foods such as maize and rice within the household has been a critical source of brans and broken grains suitable for feeding livestock in traditional systems. Commercialisation of food production that leads to processing grain off-farm leads to losses in opportunities for small-scale livestock, unless a concomitant increase in fodder crop production occurs.

Recent research with farmers has identified promising strategies for breaking this dependence by using feed inputs available on-farm such as sugar cane as energy sources. High quality forages such as mulberry, hibiscus and cassava leaves can provide some or all of the protein needed in tropical livestock systems but for monogastrics suitable protein sources may ultimately constrain such systems. Soybean fodder and small fish have been used successfully for pig production in the few cases in which these are available. The ability of ducks to utilise low-fibrous plant material such as duckweed has also been shown.

Small-scale farmers may have a continued relative advantage in the production of ruminants and village poultry because of their forage-feeding base and special characteristics. Non-conventional livestock that can be reared to substitute for forest or bush meat may fill a similar niche that is expanding in many places where industrial food production is becoming the norm. In general, however, these systems must be upgraded to a level of productivity that meets the farmers' rising expectations. Clearly, some degree of intensification of livestock is necessary for any integration with fish culture to be sustainable. Market development and veterinary support, however, are often critical to the development of such improved systems.

Collection of Monogastric Wastes

The adoption of semi-feedlot systems with controlled grazing/browsing for part of the day and confinement with supplementary feeding for the remainder of the day and night-time may have the greatest potential for upgrading livestock production and allowing integration with aquaculture by small-holders. This combines the low-risk advantages of traditional systems with enhanced productivity producing more and better quality livestock waste that can be collected for use in fish culture. Aquaculture can, therefore, become a focus for improving overall nutrient use on-farm. The nutrient-poor status of small-scale farms contrasts with the high losses of nutrients, particularly N, which occurs before collection and reuse. The N from backyard poultry waste and urine from ruminants is often almost completely lost. Whole-farm strategies that incorporate fish culture may enhance conservation and recycling of these nutrients, particularly if livestock are raised in improved, semi-feedlot systems. Such systems may also favour the strategic import and use of nutrients as feeds or fertilisers on to the farm. However, substantial and cost-effective improvements in productivity of the whole system probably require both better recycling and use of more inputs.

A major problem with upgrading scavenging poultry flocks to a size that their manure can have an impact on fish production is poor survival of starter birds, and poor availability and/or cost of supplementary feed for growing birds. The strategic use of higher quality feed for starters and use of alternative supplementary feed for adults can increase flock size, and quantities and qualities of waste available.

The adoption of heat stable vaccines to prevent Newcastle's disease has had a major impact in areas where available. The collection of waste from scavenging poultry overnight is often possible even when they remain unconfined; roosting of birds over ponds can be encouraged with strategic feeding and placement of perches/ cover. Manure can even be collected under tree branches or within housing by placing sheeting under roosting positions and feeding stations. The sheets, which can be old textile material or split fertiliser bags can then be washed or suspended in the fishpond.

Ruminant Systems

Inadequate feeds and their poor utilisation are key constraints to improved ruminant systems. The greater use of crop by-products, undergrowth of tree crops and improved fodders, particularly legumes, have been promoted to, and adopted by, farmers. The three strata forage system in which grasses and ground legumes (first stratum), shrub legumes (second stratum) and fodder tress (third stratum) allows good quality feed to be produced year-round in dryland farming areas of Bali. The introduction of multitiered feeding and polyculture in these terrestrial agroecosystems is analogous to the multiple feed/ space niches typical of fertilised fish ponds stocked with a polyculture of fish.

The use of supplementary concentrates such as ricebran can further improve performance, especially of pregnant and lactating animals. The biggest challenge is to improve the feeding value of the large amounts of fibre (ligno-cellulose) available. The use of pretreatment using alkali or urea, which improves fibrous feed quality, has been promoted but is not widely adopted by farmers, except in China where it has been successful. Although the use of multinutritional blocks, providing nitrogen and micro-nutrients, are being used in several countries to improve the utilisation of low quality feeds, in practice their use is often constrained by their poor availability to farmers in rural areas.

Improved feeding often develops with increased confinement of ruminants on smallscale farms in Asia. Animals that are stalled some or all of the time, and fed cut-and-carried forages and/or concentrates, tend to show increased productivity and produce wastes of higher quality and collectability. Both low nutrient density and high concentration of tannins negatively affect water quality and limit the amount of faecal wastes that can be used in fish culture. An optimal strategy under many conditions would be to use solid wastes to fertilise crops and to collect urine and washing water for fish culture. Normally, collection and retention of urine N in the farming system is problematic. Generally the urine simply soaks away or is absorbed by bedding and solid faecal waste. Use of greater quantities of bedding materials high in carbon (C), such as straw or stover, improves N conservation but these may be unavailable in the quantities required. Nitrogen is also lost or becomes refractory under such conditions. Pens can be constructed with a thin sloping concrete floor to direct urine and washing water into a plastic bag. Where concrete is unavailable, elevated wooden pens, which are often used for stalling small ruminants, can be sited directly over the pond. In practice, however, livestock are often penned away from the pond and waste has to be collected and carried. There is a need to develop practical methods to collect urine, perhaps using simple plastic sheeting and bags.

Mixed Input Systems

In many situations livestock wastes will be only one of several nutrient inputs used to raise fish. Livestock wastes may be too few, low quality or available irregularly. Multiple inputs are used in both subsistence and commercial fish culture with livestock waste to capitalise on an occasional or seasonal abundance of other wastes and byproducts and to improve the overall balance of inputs into the pond. The use of multiple inputs usually suggests that livestock production does not dominate fish economically. Use of livestock manure alone usually suggest that aquaculture is a relatively minor component of the overall system.

Smallholders

Various factors may limit the number of livestock that a farmer can manage and integrate with fish culture, reducing wastes to levels below optimum. Edwards et al. found that problems marketing duck eggs, and high feed costs, constrained small-scale farmers maintaining even 30 ducks over small ponds as feedlots. Farmers with limited numbers of poultry for their pond area need additional nutrient inputs to optimise productivity. Extension of multiple input systems can increase overall levels of inputs used by small-scale farmers, although often not to the levels considered optimal by researchers. In a project in Bangladesh the frequency of poultry waste used as pond inputs increased from only 3 percent of farmers to nearly 30 percent after extension. Comparative figures for cattle manure, inorganic fertiliser and rice bran reflected higher initial use.

Supplementing low-nutrient density wastes from scavenging animals with other inputs is particularly common. Farmers in Udorn Thani, Thailand used a variety of inputs, in addition to livestock manures including a variety of plant leaves, and rice bran. The wastes collected from limited numbers of scavenging ducks fed a supplement at night can increase fish yields in systems fertilised inorganically by over 40 percent. Additionally, the overall efficiency of the limited ricebran used on the farm is improved and products diversified. In small ponds, the relative amounts required are also affordable, given the value of the fish produced.

Commercial Systems

Commercial, integrated livestock-fish systems are best developed in countries such as China, Taiwan and Thailand where vertically integrated feed companies have exploited rapidly growing markets for meat and social, market and physical conditions are optimal for using the large amounts of waste through semi-intensive fish culture. Increasingly these are complex, multi-input systems in which the use of formulated feeds, aeration and improved strains are being adopted to increase yields and returns. The evolution of integrated fish culture in China gives a particularly interesting profile of how heterogeneous conditions stimulate different practices.

Several studies based on data in the 1980s do indicate some important trends in the country's transformation.

In locations where market and physical conditions are suitable for both livestock and semi-intensive fish production and the concentration of livestock production lowers the opportunity cost of the wastes such as Chacheongsao and Nakon Pathom provinces near Bangkok in Thailand, direct physical integration of livestock and fish production develop as in livestock dense provinces of China. Fluctuating livestock prices are common under such conditions and Edwards showed that fish production makes a significant contribution in covering losses when livestock prices are low.

The trends towards simplification of integrated systems with single species of livestock raised at high density, and the use of supplementary feed and aeration has developed over the last two decades in China. There has also been substitution of livestock-fish integration with more feed-based, fish production and diversification to higher value species. This appears to be mainly occurring in areas close to the major, richer markets where land is most limited, fish prices are high and high quality feeds are available. Nationally the situation still reflects dependence, and increased growth, on fish raised in fertilised ponds and it is clear that China will need to depend on integrated concepts to sustain inland aquaculture.

Farmers with limited numbers of livestock may use fertilisation based on both livestock wastes and inorganic fertilisers. Inorganic fertilisers may be a cheaper form of N and P than purchased livestock manure in many situations, and highest yields may be achieved without, or with relatively low loadings of manures. This may be particularly the case for phytophagous fish such as Nile tilapia. In highly productive ponds fertilised with inorganics, there may be no significant benefit of including chicken manure in tilapia monocultures.

The benefits of using high levels of inorganics may be constrained by their availability or opportunity cost. Other species may benefit more from use of manures, especially when high levels of inorganic fertilisation may be less appropriate such as for carp polycultures which include a detritivore. Supplementation of organic wastes in low-input, chemically fertilised ponds increased the growth rate of mrigal, the detritivorous Indian major carp compared to controls.

Supplementary feed

Supplementing limited livestock manure with direct feeding of fish is an alternative strategy. The impact of supplementary feed on yields of fish in ponds fertilised with livestock waste is affected by many factors. The level of natural feed to some extent affects the efficiency with which fish utilise supplementary feed. More natural feed allows use of more high-energy supplements to 'spare' the protein requirements and support the growth of more fish. The optimal levels for supplementary feeding are complicated by the variable levels of waste feed that are mixed with the manure which may be considerable. Purchasing supplementary feed for fish in livestock-fish systems may not be cost effective however, especially if feeding rates are high.

A reduction in feeding costs of more intensive systems by fertilising ponds with livestock manure is another strategy that has attracted the attention of farmers and researchers alike. As feed cost typically comprises about 80 percent of operating costs of intensive aquaculture systems, any improvement in food conversion efficiency is desirable. Clearly, the fish species raised need to be suitable for culture in waste-fed, planktonrich systems. Green et al. found that the tambaqui, in contrast to the Nile tilapia, grew poorly in fertilised systems without supplementary feed. Increased levels of inputs have consequences for overall loading rates, water quality and fish yields.

Microphagous Nile tilapia fed pelleted feed in duck manured, mechanically aerated ponds in Taiwan yielded up to 18 tonnes ha-1 however in systems that produced most of the low cost tilapia exported to the USA. Hepher and Pruginin's description of commercial polycultures in Israel indicates that fertilisation with livestock waste and inorganics was an essential component of high-yielding semi-intensive systems in that country. Feeding only in the later stages of the culture period, when the nutritional needs of the fish exceed the level supported by natural feed alone may also reduce feed costs. Green et al. found that at densities of 1 fish m-2, tilapia could be raised on poultry waste alone for the first 90 days of a 137 day production cycle without any differences in final yield.

Integration with Agro-Industry

Global market integration and commercialisation greatly increase the volume of animal waste. As incomes increase, a greater proportion of animals and animal products are processed. Large quantities of homogenous wastes produced during production and processing of feedlot livestock make for economic use of such by-products for fish feeds under a variety of conditions. Re-feeding manure and slaughterhouse waste back to livestock and fish has been common in recent history.

Early accounts of trout production indicate a dependence on old horses as fresh meat or as a substrate for maggots used as feeds. However, the use of meatmeals has come under increased scrutiny recently after ineffective processing of slaughter wastes led to outbreaks of bovine spongiform encephalitis (BSE) in cattle. Fish and shrimp meals are widely used as ingredients in both animal and fish feeds. Both livestock production wastes i.e. manure and slaughterhouse wastes may be processed into conventional feed ingredients used by feed mills, although only the latter is widespread.

The most important environmental impacts of processing livestock result from the discharge of waste-water. This area is least developed with respect to integration with fish culture. A major problem in developing countries is the tendency for processing plants to be located in peri-urban areas where infrastructure and markets are best developed but land and water are limiting. Other major differences between modern and traditional agro-industry are the facilities and scale of operation. There are large differences in the type and quantities of by-products produced that depend on the type of livestock and processing waste.

Regulation of the size and distribution of processing operations can encourage the recycling of processing wastes through aquaculture. In Thailand, as road and other infrastructure have improved, chicken processing plants are moving away from flood-prone rice growing areas in provinces near Bangkok to the centres of broiler production within maize growing, upland areas. Hybrid catfish farms which utilise much of their wastes, are similarly relocating to these areas.

A major constraint to the use of most tannery waste is the use of chromium in the tanning process. Chromium is particularly toxic to fish and other aquatic organisms. It is a major problem in the Calcutta wastewater-fed fisheries, causing seasonal mortalities and undocumented risks to humans from consumption of such fish.Adding value to livestock wastes through simple on-farm processing has now developed where conditions are suitable: typically in the vicinity of larger towns and cities; and usually by entrepreneurs well connected with urban markets. In one study, livestock processing wastes in Pathum Thani, Thailand were more commonly used by farmers with diversified farming systems, than rice farmers adopting fish production. Location of aquaculture within peri-urban zones also makes the use of other agro-industrial by-products possible. These include waste human food from canteens and restaurants and food processing wastes. In Thailand and some other parts of Asia, noodle processing waste, stale noodles and bread products, and soybean, brewery and distillery wastes are commonly used.

Live-feed organisms

The use of livestock manures as substrates for livefeed organisms such as insect larvae or crustaceans for feeding high-value fish also has potential where land costs prohibit semi-intensive aquaculture. Live feeds can be used as complete diets or to supplement pelleted feeds for fish species such as hybrid Clarias catfish raised in intensive systems. Such low-tech solutions not only produce fish but also increase the value of the original waste, principally by reducing smell and moisture level.

Commercial examples of poultry slaughter house wastes being used as feed after simple processing for carnivorous fish such as Clarias catfish and snakehead exist in Asia. The rapid deterioration of viscera and other slaughterhouse wastes restricts their use to areas close to the source. In Thailand such wastes are usually refrigerated on-farm and may be traded. The advantages of an export industry that both stimulate employment producing value-added products and produces food for sale locally are clear.

4

Pastoral and Arable Farming

The settlement of arable farmers and pastoralists are spatially, economically and socially related. Although arable farmers are disinclined to herd cattle, even if they own them, they appreciate cattle food products and the manure, draught and transport benefits which cattle provide; pastoralists also appreciate the benefits of convenient markets for livestock products and access to crop residues. Many pastoralists are becoming agropastoralists by including subsistence crops in their farming systems.

Apart from seeking the same social amenities like schools and health facilities as the arable farmers, many now have the same ecological and market needs which result in their settlement within, or on the periphery of, arable farming communities. Pastoral women sell sour milk and butter along with prepared millet or sorghum which is mixed with the sour milk at the time of sale. This can bring about a partnership between the pastoral and the arable communities to trade their different products to the same consumers.

Pastoralists are generally classified according to their mobility: nomadic, semi-nomadic, semi-settled, settled. In terms of crop/livestock integration however, it is more meaningful to classify them according to their enterprise system and their degree of contact with cultivators, recognising that individual pastoral units are variable and modify their production strategies according to perceived risks and advantages. It is sufficient to distinguish between:

a. *Full-time livestock keepers*: ranging from those with no consistent association with a particular farming/ land-use system (nomads), to those who have more or less regular contact with cropping systems at their grazing sites. This group can be referred to as 'pure pastoralists'.

b. *Livestock keepers who practise some cropping*: have consistent association with a particular farming land-use system, and could exploit improved opportunities for integrated crop/ livestock production. These 'agropastoralists' can be subdivided into those who crop at one site but move all or most of their livestock to other grazing areas during the non-cropping season—transhumant agropastoralists, and those who keep livestock throughout the year near their cropping activities—sedentary agropastoralists.

Apart from better access to crop residues there are other reasons why pastoralists wish to settle and integrate their livestock with crop production:

a. The inability to survive on livestock products and revenues because of drought or disease.

b. Traditional cattle husbandry systems will become more difficult to sustain due to encroachment on grazing land by crop producers.

c. Cattle movement will be impeded by ribbon development along highways and water-courses that cut across trek routes.

d. Sanctions against cattle movement such as veterinary, quarantines or tolls.

e. The need to settle to obtain access to veterinary, livestock feed and social amenities such as education, health and welfare facilities.

f. Acquiring cattle as payment for herding cattle owned by local farmers.

These factors will vary in priority but all will depend on the availability of grazing land which will be determined by:

a. Competition for available land by cultivators.

b. The relationship between the pastoralists and their prospective neighbours.

c. The land tenure status and the degree of government intervention in the provision of land for pastoralists.

The acceptance of a sedentary life style is a major revolution in the socio-economic life of pastoralists. It affects their individual and collective self-confidence, especially when settlement is imposed.

All extensive grazing systems subject to drought must retain the ability to move livestock. Renting grazing and agistment are common features of ranching in the Americas and Australia. Unlike commercial ranchers the pastoralist has misgivings about adopting sedentary systems of production.

Agropastoral Livestock Production

There is ample evidence to show that low productivity is not wholly genetic and that local cattle breeds are capable of much higher performance given better nutrition. Nevertheless, owing to the poor quality of the natural rangeland and short daily grazing time cattle production cannot be increased without feed supplementation. Johnson et al. determined, by simulation modelling, that improved nutrition was the most important factor affecting performance; when high quality feeds are scarce the best returns come from supplementing breeding females. Given the shortages of agroindustrial by-products and other feeds, however, this supplementation will normally come from improved crop residue utilisation and forage production on fallow lands.

Agropastoral Crop Production

Although pastoralists usually settle to take up subsistence cropping they do not dispose of their herds indicating that production of subsistence food may be a means of preventing the depletion of herds that are too small to provide total household food and cash requirements. Fortunately pastoralists can usually obtain some crop land because most communities traditionally acknowledge the right to land for subsistence cropping. This will, however, become increasingly difficult as land becomes more scarce.

Pastoralists typically adopt the farming habits of the local arable farming community. The Fulani at Abet near Kaduna, northern Nigeria, have adopted the local farmers' techniques of planting millet in nurseries for transplanting into their fields; they also fertilise the nurseries with goat manure and wood ash, use similar crop rotations to control pests and diseases and purchase similar inputs like seed dressing and chemical fertiliser.

The pastoralists ranked sorghum, maize, millet and rice as the most important crops. Their willingness to adopt appropriate crop techniques indicates their capacity for change, but their basic view of arable farming as a means to produce subsistence food and their preference for investing in cattle production, influences their willingness to accept innovations.

The availability of artificial fertiliser as opposed to animal manure, especially when it is subsidised, may also impede the acceptance of crop/livestock systems. Some farmers claim that trampling by cattle makes the soil harder to work and that manure carries weed seeds, but most farmers acknowledge its effect on soil fertility; if animal manure is in short supply pastoralists can earn about US$1 per head per week for tethering cattle on farmers' fields. When manure is readily available pastoralists may have to pay to graze the crop residues in cash and/or overnight manuring.

Apart from the few who can live on grazing reserves or ranches, pastoralists need to associate with arable farmers to supplement the feed supply of their herds. While there may be good reason for localised specialisation in livestock or crop production the two systems will always be interdependent because they share the same land and rely on a mutually beneficial exchange of products.

Trypanosomiasis Risk

With the increase in the intensity of farming and hunting, there has been an overall reduction in natural tsetse habitats and hosts which has led to a decline in tsetse abundance; this occurred within areas prior to tsetse eradication and in regions outside the eradication zone. Nevertheless, riverine tsetse species are capable of persisting where suitable habitats remain, even within areas of high land use intensity, but in the absence of wildlife they have become increasingly dependent on cattle and/or man for their survival. Under these circumstances, the trypanosomiasis challenge will be diminished because of the restricted distribution and reduction of tsetse populations. In addition, pastoralist herd management practices minimise the period of direct contact between the tsetse fly and their cattle.

Where trypanosomiasis incidence has fallen because the wildlife reservoir for the disease has disappeared, a fundamental change in the balance of vector-host-disease relationship takes place, which favours the development of immune responses to trypanosome infections in cattle and results in the selection of less pathogenic trypanosomes and thus less virulent forms of the disease. Such reduction in the virulence and incidence of trypanosomiasis might be encouraged by the more sedentary life-style of the pastoralists, which has reduced the range of strains of trypanosome to which their cattle are exposed. In addition, the decline in long-distance migration has reduced the periods of stress resulting from trekking which render cattle more susceptible to infection.

Relevant Policies

As livestock development policy tends to be the responsibility of specialised agencies there are few national policies directed specifically towards the integration of crop and livestock systems. Development plans for pastoralists have tended to concentrate on infrastructure to encourage settlement rather than increasing agropastoral production; plans to assist milk producers have concentrated on the building of milk collection centres rather than promoting production per se. Forage production technology has not been adapted to pastoral circumstances.

Some integration measures like animal traction and cattle fattening on smallholdings which were developed for mixed farmers are suitable for agropastoralists. Many pastoral development schemes in grazing reserves have, however, discouraged cultivation by herd owners, and service centres and credit are presently not suited to the needs of agropastoralists.

Unlike the nomads the agropastoralists are exposed to commerce and other external influences although, apart from veterinary services, the impact of government development and advisory services has not been great. This is largely because, until recently, government officials grouped pastoralists with nomads, who were considered unreachable apart from their inclusion in mass animal health control measures. Most development projects were designed to induce settlement rather than to promote the welfare of those agropastoralists already settled, and contained large infrastructural components like roads, firebreaks, dams and milk collection in new areas set aside for settlement.

There are few explicit national policies for integrating crop and livestock production orientated to the specific needs of the pastoralists; integrated crop/livestock production involves more than draught power and smallholder fattening schemes. Further, because arable farmers form the majority of producers agricultural development projects have tended to ignore pastoralists. In view of the rapidly growing human population and the need to increase food crop production it can be assumed that all potential arable land will eventually be used for crop production. As a result there is a tendency to lower the priority, or even deny the need, to include livestock in the development of arable areas.

Whilst it is true that in a food deficit situation all suitable land should be devoted to crop production, this does not imply that ruminant livestock should be excluded from these areas because:

i. there will always be a demand and good dictary reasons for animal protein;

ii. with a single crop growing season, ruminant livestock provide the only land use opportunities for the rest of the year;

iii. there will always be residual rangeland;

iv. the poor carbon and nitrogen content of the soils could be improved by the incorporation of forage legumes into cropping and fallow systems;

v. over 50 percent of the dry matter produced by grain crops can be utilised only by ruminants;

vi. dairying and intensive fattening systems provide income and cash flow benefits that can assist in the timely purchase of agricultural inputs.

To counteract the neglect of livestock producers most governments have set up specialised livestock agencies. These specialised agencies, even if they support integrated crop/livestock development, are not adequately staffed or trained to work with arable farmers. This dichotomy is also evidenced by the separation of agricultural or food sector reviews from livestock sector reviews.

All governments have detailed policies and plans for comprehensive, but usually separate, crop and livestock extension services. The spread of new varieties and increase of fertiliser use are evidence of an extension impact on crop production. These inputs have quick and visible results and are, relatively easy to sell to the farmers; further, arable farmers are usually concentrated in small areas. Livestock extension staff, on the other hand, tend to be poorly motivated because they have insufficient contact with cattle owners and, as a result, spend much of their time on infrastructural development.

Although it is recognised that there is sufficient genetic variability in the indigenous breeds on which to base selective breeding for improved productivity, this would take a very long time to make any significant impact on national herd productivity for milk. For a rapid improvement in milk production it will be necessary to use imported dairy genes provided an adequate proportion of indigenous blood is retained in the cross to maintain natural adaptation to the environment. Introduction of crossbreeding in pastoralists' herds however should await the solution of health, nutritional and management problems.

Forage production is now receiving priority and governments and research institutions are becoming more conscious of the need to undertake adaptive or systems research. The results of past research have been disappointing in terms of uptake by agropastoralists although it has created a good technical base for future work.

A well-tried and successful method to integrate crop and livestock production is the introduction to the arable farm of animal traction. It should be noted, however, that when pastoralists take up arable farming they do not normally exploit their cattle experience by utilising draught animals. If the local agricultural community does not use draught animals then neither will the pastoralists who settle within it. There is a need for extension and training in the use of draught animals where these can be employed efficiently.

The easiest innovation to offer to both farmers and agropastoralists is the stall fattening of a few cattle. It has the advantage, in a relatively short time, of increasing saleable beef production, producing better finished carcasses, stimulating off-take of surplus males at a younger age, utilising crop by-products and introducing the agropastoralists to commerce and credit.

Grazing reserves and group ranches have been the focus of development plans because of the lack of alternatives, and because their proposers knew that, with the spread of arable agriculture, there was only limited time to allocate land on a large scale to livestock producers; to date, however, the number of ranches demarcated is less than the number planned.

Pastoral development plans often include projects for parastatal and/or private ranching. These should be viable as commercial production units or designed to handle special functions such as the importation of foundation trypanotolerant stock. These ranches are often expected to provide services for the traditional sector: a sale venue for immature stock, a source of improved breeding animals and a demonstration of the advantages of modern management. The impact of this secondary role has been disappointing; in particular they have not demonstrated crop/livestock interactions.

Many countries are developing service centres where advice and inputs are made available to farmers and where extension staff can make contact with producers. Any serious attempt to integrate crops and livestock will require coordination at both political and field level. Equipping agricultural service centres to supply the needs of the livestock producers would be a major step to achieve this coordination.

Government policy towards credit for livestock development has centred around the provision of supervised loans to producers. Because the specialised agricultural credit agencies do not have adequate branch offices with supervisory staff, loan supervision has become the responsibility of extension staff. This indirect supervision has not encouraged adventurous development of loan portfolios in either quantity of funds or range of activities. At the same time project planners continue to apply pressure on reluctant credit institutions to accept loan programmes to finance the cost-recoverable portions of their projects.

Livestock production is constrained over vast areas by the presence of the trypanosomiasis-carrying tsetse fly and a great deal of investment has, and is continuing to be, made in tsetse eradication. Although tsetse clearance is not a prerequisite for settlement in the subhumid zone it does speed up direct settlement but, if clearance is done haphazardly without adequate post-eradication land use planning, much of its value will be lost. Nevertheless, there are few guidelines on how land should be used once it is free of tsetse. These policies and guidelines should be directed towards integrated crop/livestock systems or pastoralists will be excluded from large areas of the subhumid zone.

Impediments to Integration

The major constraint to animal production is malnutrition. In the arid zones this is due to overstocking; in the humid zones the poor quality of the available forage is responsible.

Crop production systems that require long periods of fallow are under pressure due to shortage of land and labour. Land tenure is complex and uncertain because the agropastoralists are usually minority, non-landowning communities. Grazing reserves and group ranches are not fully effective because they tend to be designed in isolation from overall land use systems and even internally do little to promote integrated crop/ livestock production. The lack of routine, developmental inputs and inappropriate credit mechanisms are major impediments to the development of integrated crop/ livestock systems.

Lack of Adequate Research

Although the majority of pastoralists have identified their own time and place to settle conferences and planners persist in trying to develop ideas and hypotheses on how to encourage them to settle. Much less attention has been given to assisting those that have already settled. This lack of understanding of the pastoralists is in part due to the fact that most socio-anthropological work has been done amongst nomads who principally seek water and grazing.

Reliance on reshaping old remedies will continue until the research stations can collect applicable field information on farming systems. There needs to be active cooperation between research station and field scientists and between scientists of all disciplines.

There is still no adequate livestock systems research being carried out in any research institute, although there is almost universal support for the concept. There is need to define who should carry out the work, what resources should be committed to it, how the teams should select topics and how feed-back mechanisms should function between systems teams, the research stations and development agencies. There is a tendency for research staff to be overly critical and for extension staff to be defensive concerning the promotion of new techniques.

Lack of systems research information about arable farming has meant that multiple-cropping has only now become accepted as a recommendable practice after many years of promoting sole cropping, usually against the experience and wishes of the beneficiaries. Even now planners and advisers know little about the efficiency of the multiple cropping systems which form the existing production base. The projects thus tend to concentrate on infrastructural and input supply systems, while the farmers' ability to utilise the improved opportunities is constrained by their traditional practices.

Animal Diseases

Although major epidemic diseases such as rinderpest and contagious bovine pleuropneumonia still constitute a serious threat to cattle, this situation is due to failure to maintain adequate vaccine production and application rather than to lack of efficient vaccines or of knowledge of the epidemiology of these diseases. Streptothricosis may cause losses particularly among exotic breeds and their crosses but experience suggests that efficient tick control measures go a long way toward controlling the incidence of this disease.

The adoption of more intensive management systems could result in some shift in the herd disease spectrum with better control of epidemic disease becoming facilitated as herds become more static and therefore more accessible. Greater contact frequency between animals could bring higher incidence of bacterial diseases such as tuberculosis, brucellosis and gastro-intestinal parasitism. However, malnutrition is now the main limiting factor in herd productivity, and is likely to remain so until such time as supplementation of natural range and better utilisation of agricultural by-products have become successfully established.

Shortage of Labour

The greatest impediments to increased crop production by agropastoral households are shortage of labour and access to additional arable land. The latter may be overcome through the more intensive use of smaller plots but this will lead to other problems as the present agricultural systems are not designed to produce long-term returns from the same piece of land. The agropastoralists have access to animal manure but in most instances, outside grazing reserves and group ranches, they provide this manure to the cultivators with whom they associate.

Lack of Security

Land reform and some form of land title can offer advantages through physically, and formally, linking particular persons to certain pieces of land; this has great appeal to farm planners and bankers and has been carried out in Kenya with some success. It is unlikely, however, that any government would be prepared to undertake any large land reform programme with the result that there is some doubt that any satisfactory solution can be found to underwrite long-term land improvements. Lack of land title has not prevented progress in crop husbandry with its annual cycles and traditional rights of occupancy and land use, but it has delayed progress in rangeland improvement and forage production.

The semi-arid zone has been increasingly occupied over several centuries. As the population has grown, pressure on land has intensified with the result that the typical unit is a smallholding. Little change can be expected. The humid zones present a different picture because they can still accommodate planned development. Nearly all the land, however, even in unoccupied tsetse-infested areas, is claimed by one ethnic group or another, who wish to protect their land rights for the time when their population exceeds the human support capacity of their present holdings. They are also anxious to protect their traditional procedures for allocating land and settling land disputes. Abuse of these procedures can quickly arouse hostile reactions and create an atmosphere in which progress is impossible.

Flaws in Grazing Reserves

Despite the inclusion of grazing reserves and group ranches in official plans, demarcating large pieces of land for the exclusive use of pastoralists is fraught by competing land claims. The case for the establishment of grazing reserves has not been helped by the failure to plan them in the context of total land use systems. Grazing reserve and group ranch plans typically envisage self-contained, year-round grazing systems and result in the allocation of large areas of land to few people.

Shortage of Inputs and Credit

Apart from land recently cleared of tsetse those areas which have been avoided by cultivators are unlikely to be attractive to agropastoral settlers. Although the agropastoralists' first needs are for veterinary services and food for their cattle, and seeds and fertiliser for their crops, the extension staff they usually encounter are the veterinary auxiliaries who know how to inject vaccines and dose with anthelmintics but cannot diagnose nor prescribe, and junior range management specialists who cannot advise on cropping or forage production. This situation leads to mutual frustration and minimal contact between extension staff and producers.

This frustration is exacerbated by shortage of inputs and credits. Even if inputs like salt, feedstuffs, drugs, fencing materials, seeds and fertilisers are available, supplies are sporadic and expensive. Rural retailers are disinclined to stock such goods because of the high transportation and storage costs and the slow turnover resulting from the seasonality of demand; neither do they have the necessary cash nor incentive to finance such inventories. Although the pastoralists may be aware of the benefits of employing these inputs they do not have the means to finance their purchase; the daily milk revenue belongs to their wives and is used to purchase grain and household requirements. A prime need of farmers for supplier-credit, but the credit available to them is usually in the form of complex development loans with onerous conditions and supervision.

Current Agricultural Production Systems

The existing agricultural or farming systems in different ecological zones are designed to produce subsistence food, cash sales and materials for local or industrial use. The various production systems are grouped under:

(i) traditional and transitional systems, and

(ii) modern systems and their local adaptations.

Traditional and Transitional Agricultural Systems

The majority of the farming systems belong to this group which range from the extensive (shifting cultivation and nomadic herding) to more permanent and specialised types of farming (compound farms and terrace farming). Shifting cultivation is an extensive agricultural production system in which a cultivation phase on 'slash-and-burn' cleared land alternates with a fallow period. The clearing is done using axes or matchetes and usually only herbaceous plants, saplings and undergrowth are cut. When dry the cut material is burned and the cleared area is planted on the flat or on mounds with crops like yams, rice, sorghum, millet, maize and cassava depending on the ecological zone. The land is cultivated for one to four years after which it returns to fallow. The regrowth of natural vegetation rejuvenates the soil through nutrient cycling, addition of litter and suppression of weeds. The cultivation phase alternates with a much longer fallow phase, ideally 10-20 years.

In classical shifting cultivation, the homestead of the farmer is relocated near the cultivated fields after each cultivation phase. In practice the situation varies from where the farmer may never return to the same piece of land to situations where cultivation is repeated on the same plots. This cultivation system is ecologically viable in frontier situations where population density is low and fallow periods are long enough to restore soil fertility. Nomadic herding is the extensive animal rearing counterpart of shifting cultivation in the savannah and more arid areas is discussed separately.

During the colonial era trade in spices, forest products and later cash crops led to increased cash cropping and with increasing industrialisation, the production of export orientated raw materials. With increasing domestic industrialisation many of these raw materials can be sold locally. With increasing urbanisation and mobility, the market economy has become well established, particularly the need by farmers for money to purchase goods, services and farm inputs which can only be obtained through sales of farm produce. By products of farming can also be used. Therefore the objectives of traditional farmers are various and, although food crops still have priority, some commodities are produced specifically for sale.

Spatial arrangement and structure

Traditional farming systems consist of several fields in more or less concentric circles round a compound or homestead-garden. The homestead-garden is under intensive, permanent production and contains a mix of perennial and annual crops grown in a complex agroecosystem. In the tropical rain forest zone this may attain a multistoried structure approaching that of the tropical rain forest. In the savannah, it is not so complex owing to lower rainfall but the diversity index usually exceeds that of the surrounding woodland. Fertility is maintained by applying household refuse, crop residues and animal manure. The fields are located at varying distances from the homestead and have fallow periods which increase with increasing distance from the homestead.

There are specialised cropping systems that take advantage of topographical or microenvironmental variations which occur within the farm. The homestead-garden is usually sited in a dry, upland situation, but nearby may be lowland or valley bottoms where rice, vegetables, yams, sugar cane, bananas, taro, etc. may be grown depending on the ecological zone. Crops which require high levels of fertility or greater care like yams, vegetables and cocoyam in the forest zone or tobacco and vegetables in the savannah, are grown nearest to the homestead. The homestead-garden system includes varying numbers of small ruminants, pigs and poultry; cattle, donkeys, camels and horses are kept in tsetse-free savannah areas. Livestock are a feature of the homestead and adjacent areas and they graze on fallows and residues on the harvested fields.

The different production systems compete for labour and resources and to maximise these, different crops or varieties of a crop are grown in the different fields. With distance from the homestead, fishing, hunting and some food gathering increase in importance. This concentric field system is sometimes called 'land zoning', or the 'ring system of cultivation'.

Farms and production methods

Since the early 1930s use of animal power for cultivation, planting and transportation on the farm has been increasing in areas free from tsetse. Use of tractors and tractor-drawn implements is increasing but is of no general significance. Land is almost always universally cleared by manual cutting and burning. Thereafter there is little or no tillage and crops are grown on the flat, or on mounds or ridges which are manually constructed. Weeding is usually carried out manually. There is increasing mechanical land clearing as a result of the high cost of labour.

Agriculture in developing countries is characterised by a division of labour between the sexes, with women specialising is some operations and the men in others. There is often acute shortage of labour on farms due largely to seasonal demand peaks for farm operations and division of labour between sexes and age groups. Shortage of labour is aggravated by rural-urban migration, children attending schools and competition from the non-agricultural sector.

Apart from organic and animal manures very little use is made of chemical fertilisers. Physical and cultural methods are used to control pests and diseases although pesticides are used on some cash crops. Traditional farming systems are designed by the timing of operations, species grown and resource manipulation to make best use of the prevailing rainfall regime. Although there is some traditional use of water to supplement crop growth very limited use is made of irrigation.

Crops are usually grown as mixtures, relay intercrops or associated sequences. Crop rotations and sole crops are rare apart from rice and to some extent cash crops like groundnuts, cotton and sugar cane; the more a crop is grown for sale the greater the likelihood of its being grown in pure culture.

While productivity per unit of energy may be high, yields per unit of area are low in traditional farming systems. There is not only a wide gap between attainable yields on farms as compared to experimental stations but also as compared to average and recorded yields in other parts of the world.

Traditional and transitorial farming systems in developing countries are complex and not only involve several species of crop but often include livestock. Although crop production is not a component of nomadic herding there are many examples of complementary and symbiotic interrelationships between farmers and herders.

Traditional farmers, their wives and farm family members in rural areas usually engage in many paid non-farm activities. Earnings from these activities significantly contribute to family income.

Changes in Traditional Farming Systems

Farming systems have reacted to changing circumstances. Changes have resulted from

- (i) introduction of new world crops;
- (ii) population expansion;
- (iii) European colonisation of Africa and the need for spices and agricultural raw materials for industry;
- (iv) improved means of transportation and communication;
- (v) expansion of cassava production into marginal areas where other crops often fail, and
- (vi) introduction of mechanisation into farming and adoption of new techniques.

Modern farming systems and their local adaptations often exist side by side with traditional systems.

Towards Integrated Crop and Livestock Production

The first system of crop production which replaced hunting and gathering was shifting cultivation. It was an efficient and reliable production system in situations where population density was low. As population density increased, however, fallow periods became shorter and more intensive and semi-sedentary agricultural production systems replaced shifting cultivation. Sedentary agriculture involved settlement in villages and communities in which the homestead became a permanent feature associated with a homestead-garden. The outlying fields associated with each homestead were subjected to different periods of cropping but, when population became very dense, these fields were permanently cropped. It would appear that in almost all parts of the world traditional farming systems centred around a homestead-garden where livestock were kept.

Continued intensification of production often resulted in more specialised production systems which required more inputs, more capital, specialised practices and new tools and technology. Animals rarely feature in the specialised crop production systems, except where mixed farming is possible, but animal populations increase with increasing human population density. The animal component of the traditional farms plays a crucial role in the maintenance of soil fertility, without which the permanent farming system used on the homestead-garden and adjacent areas would be impossible. Livestock provide meat and other animal products, form a savings account, utilise household waste and fulfil social and cultural obligations.

Policy Initiatives

In almost all countries national policies in agricultural development have changed very little with respect to integration of crop production and livestock rearing. There has, however, been a shift of emphasis since independence. During the colonial era, priority was given to research, extension, marketing and infrastructural development that enhanced production of cash or export crops to the detriment of the major staples. Since independence many African countries have experienced rising food import bills at a time when they are already having balance of payment problems with high petroleum prices and/or the result of adverse effects of drought. Consequently, emphasis has been shifted from cash crops to food crops. Although during the colonial era higher priority was given to crops than to livestock, considerable efforts were made to control major livestock diseases like rinderpest, bovine pleuropneumonia, trypanosomiasis,

East Coast fever, anthrax, contagious abortion and parasites such as helminths and ticks. The alarming losses caused by these pests and diseases reached such a magnitude that greater emphasis was given to veterinary services than to the equally important animal husbandry practices and nutrition. Veterinarians dominated policy concerned with animal production and, even when many of these diseases had been brought under control, low productivity attributable to nutrition and husbandry remained neglected. Sometimes animal breeding and production were grouped into a separate Ministry while crop production, which should include forage and pasture crops, belonged to the Ministry of Agriculture. Thus integration of crop husbandry with livestock production was as difficult as integration of veterinary services with animal husbandry, even though animal health and husbandry were very closely related.

Since independence many agricultural production projects involving government corporations, parastatals, state farms and farm settlement schemes have been implemented; most have been unsuccessful. Very few of these projects were aimed at the smallholder who produced most of the food crops and also some of the cash crops. During the last decade, however, special projects aimed at increased food crop production have been launched and include the Operation Feed Yourself (OFY) in Ghana, The National Accelerated Food Production Programme (NAFPP) in Nigeria which was followed by the Operation Feed the Nation (OFN) and, in 1979, the Green Revolution Programme. Of these the most successful was the NAFPP which aimed at making Nigeria self-sufficient in the production of basic staples (rice, maize, sorghum, millet, cassava and wheat) by (i) encouraging farmers to use fertilisers through a 75 percent subsidy; (ii) improved seed; (iii) construction of agro-service centres for input distribution, processing, storage and purchase of produce, and (iv) provision of loans to farmers. It also involved minikit trials which provided the farmers with the opportunity of becoming involved in on-farm trials and evaluation of technology thereby significantly speeding up adoption and feed-back to research.

There are currently policy guidelines which could act as an umbrella under which effective integrated crop and livestock strategies could be employed in the development of improved alternatives to traditional agricultural production systems. The current Lagos Plan of Action (OAU, 1981) to which all Economy Community of West African States (ECOWAS) countries, as members of OAU, are signatories, provides such policy guidelines.

Pertinent provisions of the Plan which accepts ICL research and development strategies as components of a planned national and integrated natural resources management and utilisation programme are:

1. Self-reliance and self-sustainment in economic growth and development.
2. Putting science and technology at the service of African development.
3. Achievement of self-sufficiency in food production and supply.
4. Preservation, protection and improvement of the environment.
5. Reduction of reliance on export raw materials and priority to development and growth based on resources within a coordinated strategy.
6. Achievement of self-sufficiency in cereals, livestock and fish products.
7. Realistic agrarian reform as the basis for improved agricultural production.
8. Enhanced utilisation of water for irrigation based on existing and new irrigation schemes.
9. Integrated development of areas freed from the tsetse fly and improvement of trypanotolerant breeds of cattle.
10. Intensification of programmes to integrate woodlots and trees in land use and agricultural practices at village level.
11. Priority in research and development to be given to:
    — increased production of plants and animals through improved husbandry techniques;

— assessment and development of natural resources;
— alternative energy sources including use of draught animals, biogas, etc.;
— maintenance of the carrying capacity of arid lands and soil conservation;
— incorporation of agri-silviculture practices into shifting cultivation.

The above policy provisions in the Lagos Plan do not specifically mention Integrated Crop/Livestock (ICL) strategy in the development of improved farming systems, but various aspects of the Plan such as alternative energy sources and mechanisation of agriculture, achievement of self-sufficiency in cereals and livestock, integrated development of areas freed from tsetse flies and the recent interest of governments and the World Bank in integrated rural development have direct and indirect ICL implications. It is the role of scientists in various disciplines to design and conduct research and studies which confirm the beneficial effects of ICL. Priority should be given to the development of appropriate technologies and production systems which can be used by the producer on his farm, priorities which will require political commitment through sound policies and allocation of scarce but needed resources.

Impediments to Integration

Constraints to increased agricultural production by small farmers are the same that impede agricultural development and adoption of integrated crop and livestock production strategy. Many of these impediments are related to the same factors which differentiate various farming systems.

Physio-chemical factors

The most important physio-chemical constraints to increased crop and livestock production are related to climatic and other environmental factors.

While there is no danger from low temperatures, high temperatures in the surface soil or in the atmosphere close to the ground of 45°C or above, especially when associated with low humidity, can cause serious stresses that are detrimental to plant and animal life. High temperatures are injurious to seedlings and may have adverse effects on rhizobia and nitrogen fixation. They make vegetation very dry increasing the severity of burning and the spread of fires.

High temperatures cause rapid decomposition of soil organic matter with adverse effects on soil texture, nutrient retention properties and the colloidal complex of the soil and increase the rate of evapotranspiration which renders light rains ineffective, causes severe losses from dams and reservoirs and, in irrigated soils, triggers off salinity problems. Uniformly high day and night temperatures contribute to low yields as a result of relatively high losses due to respiration. High temperature associated with high humidity has adverse effects on human work performance and contributes to the poor performance of temperate animal breeds in the tropics.

Tropical rainfall is characterised by high intensity, especially close to the beginning and end of the wet season, which increases the erosion hazard on land where the soil has become exposed by forest clearing and tillage. A very serious and frequent cause of famine in the Sahel is the erratic nature of the rain and unpredictable cycles of drought. In the humid tropics, especially above latitudes 6-7°N, and further inland, one or two weeks of drought may occur after planting. Drought stress is a major cause of crop failure and with high temperatures results in marked yield reduction. Related to rainfall is the constant cloud cover in the humid tropics which reduces photosynthesis and thus productivity. Insufficient sunshine in the humid tropics results in reduced response of crops like maize to fertilisers and is one reason why maize production is becoming increasingly important in the savannah areas, provided that moisture is not limiting.

Another aspect of the light factor is photoperiodicity to which some traditional short-day plants are sensitive. The short-day sorghums take longer to mature resulting in high culm yields which provide structural materials for fencing and other purposes. As a result dwarf photoperiod-insensitive improved sorghum varieties may not be accepted by farmers.

Other physio-chemical conditions are related to soils which are highly weathered, acid and of low inherent fertility. These soils are not only fragile but in areas of Guinea, Liberia and Sierra Leone are subject to laterisation; under continuous cultivation these soils develop multiple nutrient deficiencies and toxicities, but regular liming may also cause nutrient imbalance and toxicities.

The infertility of tropical soils and climatic constraints contribute to the low nutritive value and poor quality of forage. Although the highly weathered soils of the tropics possess good physical characteristics for arable farming, it is these same properties which render them more erodible. Further, the adverse environmental conditions in tropical soils which make large areas marginal for plant growth require large quantities of fertilisers, which may not be economic, to sustain high yields.

Biological constraints

Biological constraints are related to the favourable conditions that high temperatures have on pests, diseases and parasites. The extent of the adverse environmental problems (weeds, insects, locusts, trypanosomiasis, etc.) which constrain agricultural production, and various health hazards that affect agricultural development in the tropics have been discussed by Kamark.

Human diseases and parasites that affect settlement patterns and productivity include malaria, sleeping sickness, river blindness, onchocerciasis, filariasis, bilharzia (schistosomiasis), meningitis, etc., have rendered parts of the Middle Belt uninhabited and unutilised. There are also animal diseases like trypanosomiasis, rinderpest, foot and mouth disease, theileriasis, East Coast fever, dermatophilosis (streptothricosis) and parasites in both humid and arid areas which make livestock production hazardous.

Serious diseases and pests of crops include Swollen Shoot of cocoa, maize rust (*Puccinia polysoria*), cassava mosaic, cassava bacterial blight (*Xanthomonas manihotis*), cassava mealybug (*Phenacoccus manihoti*), green spider mite (*Mononychellus tanajoa*), sweet potato weevil (*Cyles puncticollis*), maize streak virus, etc. Weeds cause serious losses in the tropics and rampant weed growth is one reason why farmers practise shifting cultivation. As a result the breeding for resistance to pests and diseases and adaptation to environmental stresses are key objectives in plant and animal improvement.

Technological problems

Despite the relative antiquity of indigenous agricultural production systems, the developing countries have not yet developed the long-term scientific tradition and quality of research necessary to maintain satisfactory growth in technology and its transfer to produce rapid agricultural and economic development. At the time of independence there were, for example, interterritorial research organisations for most major cash crops and food crops but all have been closed down. The exodus of expatriate staff after independence, except for countries like Ivory Coast, weakened the research effort and retarded progress. There is still a shortage of trained manpower.

Wherever research has been carried out it has given priority to specialised individual commodity production systems and the solution of problem along separate lines. As a result the complex food production systems which require an interdisciplinary approach have been neglected. Research on food crops did not benefit from local knowledge of traditional farming systems or from the socio-economic environment within which they operate until farming systems research programmes were started at the international and some national agricultural research centres.

It is only by using an integrated approach to research and understanding the various interacting components of crop and livestock subsystems that effective priorities and programmes can be identified to establish the potential and limits of ICL production systems. The lack of this orientation has resulted in the unsuccessful attempts to transfer successful technologies from temperate countries and to establish large-scale mechanised farms in the tropics.

Thus lack of understanding of the traditional farming systems, in particular the role of livestock, led to over-emphasis on the improvement of indigenous trypanotolerant cattle for milk production and the abandonment, until recently, of the work with these valuable breeds. Lack of knowledge of the role of small animals in traditional production systems led to a wrong strategy in poultry improvement for small farmers which advocated intensive production systems which could only be adopted by the affluent, mainly because there was no economic technology to produce maize and concentrates locally.

Socio-economic Problems

Several socio-economic constraints bedevil agricultural development and some of these have hampered exploitation of the existing potential and benefits of ICL. These include:

1. Lack of the basic statistical data and knowledge of the socio-economic background, the environment, the resources and the management capacities within traditional production systems which can be used as a basis for planning, identifying constraints and priorities and the selection of strategies in research and development in agriculture.
2. Rapid population growth which, in addition to associated socioeconomic pressures, has outmoded the traditional farming systems to the extent that their underlying scientific base and the economic viability cannot now be understood and appreciated.
3. The complexity of traditional agricultural production systems, the competition between subsystems and nonfarm components, and the farmers' interest in diversification and the reduction of risk through the achievement of reasonable but sustained yields rather than maximisation of the production of one or two commodities.

4. Land tenure and associated problems of fragmentation of holdings, small farm size, the problems of economies of scale and their effects on the profitability and acceptance of certain technologies like communal land ownership which, although eliminating landlessness, deterred investment in land improvement.
5. Continuing reliance on simple farm tools and manual labour which accentuates the drudgery of farm work.
6. Shortage of labour at seasonal peaks of demand due to division of labour between the sexes, relatively low return to agricultural work, rural-urban migration and education for children. As a result farming is left to old men and women with low productivity. Despite this there is a marked underemployment during certain times in the year.
7. Lack of capital and credit for the purchase of inputs, including draught animals in areas where they can be used, associated with the relatively high cost of fertilisers, pesticides and equipment.
8. Poor extension services and the inappropriateness of new technologies resulting in a very wide gap between experiment station yields and those on farmers' fields.
9. Conflicting ideologies about development and insufficient understanding of the socio-economic parameters which control the extent and nature of government involvement in production.
10. Poor marketing and pricing policies for agricultural products which reduce the incentives that are needed to motivate increased agricultural production. Related to this is the shortage of

infrastructure which is required to allow farmers access to inputs, services and markets.

11. Poor communication among technicians, policy-makers and politicians which results in deficient policy formulation and inadequate allocation of the resources required to achieve national priorities.
12. Inadequate coordination between research and development policies and lack of integrated programmed which take into account the common resources and environmental endowments.

Farming Systems Research

The most significant development in agricultural research within the last two decades has been the formation of international agricultural research centres and the incorporation of Farming Systems Research (FSR) into research programmes. The purpose of FSR is the study of the farmers' environment as well as the various integrated components of the prevailing farming systems. An understanding of the farming systems enables them to be evaluated, modified and redesigned to fit the needs of the farmer. FSR adopts a multidisciplinary approach to research that gives priority to:

1. Baseline data collection and analysis of farming systems based largely on collection, collation and evaluation of secondary data which is then updated and made more complete by special surveys.
2. The study of existing farming systems, farmers' objectives, resources, decision-making processes, input/output relationships, sources of income and constraints to increasing productivity.
3. The analysis and understanding of thc workings of the individual components of the systems and the

constraints faced by the farmer constitute the basis for determining research priorities and strategies. This ensures that proposed new technologies and system components are relevant to the farmers' needs and circumstances and have a high probability of adoption.

4. On-farm testing of promising technologies, the evaluation and monitoring of technology adoption and wherever possible, the introduction of necessary modifications, and the identification of the farm level constraints which are responsible for the yield gap between experiment station and the farmers' fields; this also enhances feed-back to on-station research. On-farm research also facilitates the identification of the various off-farm, infrastructural and policy constraints which interact to determine the success of a farm enterprise.

FSR, because of its multidisciplinary nature and detailed study of the farmers' environment, has made agriculturalists aware of the importance of anthropological, sociological, geographical and economic literature which describes aspects of traditional farming systems ranging from the initial hunting and gathering stage to shifting cultivation and pastoralism, through various intensities of cultivation down to the numerous development projects aimed at modernising agriculture in the tropics.

The results of FSR that are pertinent to the design and development of improved cropping systems include:

1. Intercropping which has been demonstrated to give higher stable yields especially at low input levels than sole cropping. It results in more efficient use of resources and involves more uniform distribution of labour throughout the year.

2. Intercropping involves a wide range of crop species and different varieties of each species that differ in time of maturity and other characteristics which allow the farmer greater flexibility in management and utilisation of resources.

3. The practice of intercropping satisfies the objective of attaining diversification of production for subsistence on small farms, minimising risks of failure, and producing some commodities for sale. In some situations intercropping minimises losses from pests and diseases and provides better cover for the soil which reduces erosion.

4. Considerable progress has been made in the identification of compatible mixtures of crops such as maize/cassava, plantain/ cocoyam, yam/maize and sweet potato/pigeon peas for the humid and subhumid tropics and sorghum/millet, maize/ sorghum, sorghum/ groundnuts and sorghum/ cowpeas for the savannah areas.

5. Some progress has been made in developing rotations for humid and arid areas. At ILCA, in Ibadan, it has been shown that a viable rotation involving intercropping is maize/cassava followed by cowpeas alone or intercropped with chillies or tomato. It was also shown that cowpea is a better crop than soybean in rotation with maize since it leaves more residual nitrogen in the soil.

6. Soil fertility studies have shown that for sustained yields fertiliser application is imperative but fertiliser application is not often profitable during the first year after fallow. Some application of organic manure is not only beneficial but may enhance the effects of fertilisers.

7. A start has been made to study potential fodder species, pasture establishment and herbage productivity. The crop improvement programmes are giving priority to increased yield, better nutritional quality and other characteristics but especially resistance to disease and pests, tolerance to environmental stresses and adaptation to various production systems. Emphasis is being given to integrated pest management and various ways of reducing the costs of inputs. Areas of FSR in which significant progress has been made include land development and subsequent cropping patterns, zero tillage, live mulch, alley cropping with leguminous shrubs and evaluation of browse in agrosilvipastoral systems. These are summarised below:

Land Development

Probably the most crucial problem in the development of a permanent food production system for sustained yields is the development of appropriate techniques for land clearing, land preparation and subsequent soil management. In the humid tropics sustained agricultural production is possible when (i) chemical nutrients removed by crops or otherwise lost during cultivation are replenished; (ii) soil physical conditions are maintained at a favourable level by ensuring that adequate levels of soil organic matter are present; (iii) soil is kept constantly covered and erosion effectively controlled; (iv) soil acidity, nutrient deficiencies and toxic constituents are corrected, and (v) build-up of pests, diseases and weeds are prevented.

Bringing new land into cultivation remains the most widespread method of increasing food production.

Clearing new land is now expensive as a result of labour shortage, and many governments are resorting to mechanical clearing for large-scale farms and the use of hired tractor units for cultivation on small farms, although most of the mechanised largescale farms in the tropics have failed because of poor land development and subsequent soil management during cropping.

An evaluation of various manual, mechanical and chemical land clearing techniques has shown that clearing methods and soil management must be suited to the crops to be grown. Soil disturbance during clearing was found to cause serious erosion losses and irreversible soil degradation; manual clearing produced the least adverse effects but was slow and expensive. Chemical methods applied by spraying, ringing or trunk injection with selected chemicals such as 2-4-D or Tordon have been shown to be suitable for pasture establishment, but a shear blade with V-shaped cutting edge proved to be the most appropriate mechanical method of clearing for arable crops. It was economical and, when followed by zero tillage, erosion rates were lower and yields higher than with conventional tillage. With good plant residue management and control of weeds with herbicides, the soil is protected and there is no need for terracing or other expensive soil conservation measures on average slopes.

5

Concentrate Feeds in Livestock Production Systems

Livestock feeds provide the basic nutrients required for animal production, including energy, proteins and amino acids, and minerals, vitamins and other micro-nutrients. Feeds may be broadly classified as concentrates and roughages, depending on their composition. Concentrates are feeds that contain a high density of nutrients, usually low in crude fibre content and high in total digestible nutrients.

Concentrates may be high in energy, referred to as energy concentrates, such as cereals and milling by-products, or high in protein, with over 20% crude protein, referred to as protein concentrates. Concentrates may be fed in raw or milled forms as individual feeds, or may be blended or formulated into balanced rations for particular production purposes. Compound feeds may be mixed on-farm but are also produced by the commercial feed compounding industry.

Raw Materials for Concentrate Feeds

Raw materials for concentrate feeds are commonly classified into the following categories:

- *Cereals:* the main cereals are rice, wheat, barley, oats, rye, maize, sorghum and millet
- *Grains:* all cereals except rice
- *Course grains:* all cereals except wheat and rice
- *Food grains:* grains used for human food consumption
- *Feed grains:* unmilled grains to be used as livestock feeds
- *Milling by-products:* by-products from the milling of cereals and pulses, such as brans and pollards
- *Feed-grain substitutes:* dried roots and tubers, or by-products of crop processing such as molasses, maize gluten feed, distillery and brewery grains, dried citrus pulp etc. In some classifications, 'roots and tubers' are classed separately while the latter kinds of feeds may be regarded as 'other concentrates' or 'non-conventional concentrates'.
- Oil meals and cakes: products of oilseeds and fish after extraction of their oil component either by expeller methods or solvent-extraction methods.
- Other concentrates: other energy or protein concentrates including processed livestock products and industrial products such as urea and single-cell protein.

Non-conventional feeds and processed harvested forages: these include a variety of feeds not widely used in commercial livestock diets; some may be considered as concentrate feeds after processing, such as dried lucerne leaf meal, dried cassava leaf, cassava pulp, processed pea and bean meals, sal and rubber seed meals, citrus pulp wastes and others.

Use of Concentrate Feeds in Livestock Production Systems

The primary role of concentrate feeds is to provide concentrated sources of necessary nutrients for livestock production. These nutrients include not only macro-nutrients of energy and protein but also important specific nutrients such as amino acids, fatty acids, enzymes, vitamins, minerals and others.

Both monogastric and ruminant livestock are fed concentrates. Monogastric livestock have a limited capacity to digest fibre and thus require diets of higher nutrient density with a higher proportion of concentrate feeds and less roughage, particularly for intensive poultry and pig production systems. Less intensive monogastric production systems may utilise smaller amounts of conventional concentrates to supplement available feeds in the form of wastes or by-products of human food preparation, together with some forage feeds.

Many ruminant production systems also use concentrate feeds. In intensive systems, especially for milk production and in fattening periods for meat production, concentrates may form a high proportion of diets. Ruminant production systems of lower intensity sometimes utilise concentrate feeds to supplement roughage based diets, either to mitigate the seasonality of forage supplies, or to meet the higher nutrient requirements for particular classes of stock or animals in critical physiological states, such as lactating cows, weaner calves or draught animals.

Two specific additional consequences of concentrate feeding are important to note in the context of environmental impacts of feed utilisation. The first is that improvements in livestock productivity are generally associated with improved biological efficiency of feed use; the second is the role of concentrate feeds in promoting better utilisation of roughage feeds for ruminant livestock.

Livestock Productivity

The consequences of increased livestock productivity are that the total outputs of livestock products may be increased, or that required production levels may be achieved from fewer animals. In either case, the efficiency of production is improved, requiring less total feed per unit of livestock product.

Improved efficiency of feed-use is a consequence of an increase in the proportion of feeds going to useful production as production increases, and thus a reduction in the proportion going to maintenance of the animal. Improved individual animal productivity also leads to changes in livestock numbers and population structures required to maintain particular levels of production. Thus, fewer breeding animals and replacements are required due to longer productive lives, improved reproductive and survival rates, and shorter rearing periods. Increases in the overall efficiency of feed-use in production systems through these changes can be very substantial.

Improvements in productivity and feed efficiency through the improved use of concentrate feeds can be achieved in all livestock production systems. In systems already based solely on concentrate feeds, useful improvements may still be derived from improved quality and better balancing of nutrients to requirements. In less intensive systems, productivity and efficiency can be improved several fold.

Utilisation of Roughage Feeds

The feeding of small amounts of concentrate feeds can increase the utilisation of roughage feeds by ruminant livestock in some feeding systems. Improvements in the digestibility of roughages are due to the provision of necessary nutrients to promote rumen fermentation, resulting in increased fibre digestion and intakes of roughages, reduced wastes from unconsumed and undigested feeds, and increased animal productivity and efficiency.

These effects can be particularly important in feeding systems based on relatively poor quality roughages, such as in many small-holder and semi-intensive ruminant systems. It may be noted, however, that at higher levels of feeding with better quality roughages, and with cereals and/or high-sugar concentrates, substitution effects occur so that fibre digestion and roughage intakes may be reduced by increased concentrate feeding. This occurs particularly in intensive dairy production systems, with the consequences that increases in productivity in these systems may result in an increased demand for concentrate feeds.

Roles of Concentrates in Feeding Systems

Concentrate feeds have varying roles in feeding systems for ruminant animals, depending on the roughage or forage resources available and the required levels of productivity. The major roughage and/or forage resources may be distinguished as follows: Systems based on straws and agro-industrial by-products:

- Fibrous crop residues (stovers and straws)
- Fibrous agro-industrial by-products (eg bagasse, palm pressed fibre, cocoa pods, rice hulls, pineapple wastes, sisal bagasse and pulp)
- Sugar cane processing by-products (tops, pith, molasses, pressed cane stalk etc)

— Bananas and plantains

Systems based on fresh cut forages:

— Cut and carry forages (grass, legumes, tree fodders)

— Sugarcane (whole crop)

Systems based on grazing:

— Extensive grazing on unimproved native pastures

— Improved pastures with legumes

— Intensive grassland

All of these systems may require to be supplemented with different types of supplements to provide particular nutrients to improve feed utilisation and efficiency. Thus, fibrous crop residues may require additional fermentable or non-protein nitrogen, and some will require additional readily fermentable energy and nitrogen. Balanced supplies of less fermentable proteins and starches may be required, together with other necessary fatty acids, amino acids, vitamins and minerals. Seasonal changes in feed resources cause additional changes in supplementary nutrient requirements.

A variety of concentrate nutrient sources may be used to provide balanced diets. These will include traditional commodities such as cereals, brans, oil seed residues, roots and tubers, but also agro-industrial sources such as urea, molasses, fish meal, blood and bone meals, synthetic amino acids and many others. While the latter may not be used in large quantities, they may be particularly important in certain feeding systems. Concentrates vary considerably in their specific nutrient contents. The selection of concentrates to provide necessary nutrients must take account of these differences.

Livestock Production Systems utilising Concentrates

The major production systems in which concentrate feeds are consumed to a significant extent may be identified as:

— Cattle milk production
— Beef production
— Pig meat production
— Poultry meat production
— Poultry egg production

Poultry production includes chickens, ducks, geese, turkeys and other fowls, though chickens form the largest proportion of these. In addition to the above production systems, some intensive feeding of fattening sheep and goats occurs in certain areas, and buffaloes are fed in similar systems to cattle in some areas, though the overall amounts of concentrate feeds consumed by these classes of livestock are relatively small. Fish farming systems, especially for prawns, are also locally important consumers of concentrate feeds.

Within these broad classifications of species and products, however, are a great range of systems varying in intensity of production and in levels of productivity. These systems differ considerably in the types and amounts of concentrate feeds used, and in the nature and degree of environmental impacts. The different intensities of production and feed-use are described in Table 1.

Table 1. Major livestock production systems using concentrate feeds

Production system		*Types of feeding systems and feeds used*
Poultry egg production, Poultry meat production, Pig meat production:		
	Intensive;	high potential single-purpose hybrid breeds;

		Compound concentrate feeds; high quality ingredients, feeds formulated for multiple stages of production; intensive housing, usually environment controlled; high productivity, high efficiency of feed use; large scale commercial production for sale.
	Semi-intensive/ small-holder;	dual purpose indigenous or indigenous cross improved breeds; Improved traditional systems; limited straight concentrate feeds, use of wastes, by-products and non-conventional concentrate feeds (especially for pigs); roughage feeds included in pig diets; moderate to low productivity and efficiency of feed use; production for household consumption and sale of surpluses.
	Traditional (unimproved scavenging);	indigenous breeds; Scavenging feeding systems, including use of household wastes and some food processing by-products; low productivity, production for subsistence.
Cattle milk production	Intensive, zero grazing;	high potential dairy breeds; Cut fresh and/or conserved forages with balanced compound feeds or on-farm mixed concentrates; stocking densities variable (2->10 LU/ha) depending on forage feed sources on- or off-farm; rations formulated for multiple phases of production, high quality ingredients, high productivity, high efficiency of feed use; large scale commercial production.
	Intensive, grazing based;	high potential dairy breeds; Intensive cultivated pasture grazing based systems (at least seasonally), with balanced compound feeds or on-farm mixed concentrates, stocking densities commonly 2-5 LU/ha, rations formulated for multiple phases of production; preserved forage

feeding in non-grazing season; high quality ingredients, high productivity, high efficiency of feed use; large scale commercial production.

Semi-intensive; grazing or stall feeding, moderate potential dairy or dual-purpose dairy and beef breeds;

Cultivated and improved natural grassland grazing or cut fresh or conserved forage systems, stocking densities typically <2 LU/ha, with supplementary concentrates (sometimes compounded or mixed straight concentrates); locally available and on-farm ingredients, greater reliance on available cereal milling and crop processing by-products, unconventional concentrates and lower quality feeds; moderate productivity, production for commercial trade.

Small-holder traditional; mixed grazing and stall-feeding, unimproved multi-purpose breeds;

Natural grasslands and crop residue grazing, supplementary stall-fed forage feeding systems, concentrates only in form of household wastes and by-products, or unconventional concentrates, low productivity, production for household use and local trading of surpluses.

Beef production:

Intensive calf rearing; improved dairy breeds and beef x dairy crosses;

Milk replacer and calf rearer compound concentrate feeding systems, with minimum roughage, for veal production or for subsequent cereal based fattening; high growth rates, high efficiency of feed use.

Intensive cereal based fattening; improved beef breeds and beef x dairy crosses;

Cereal based minimum roughage feeding systems, high growth rates and efficiency of feed use; feeding period 9-12 months, slaughter at 12-15 months

Intensive grazing/forage or conserved forage based systems; improved beef breeds;	Grazing, cut fresh or conserved forage systems with balanced compound or on-farm mixed concentrates, stocking densities typically 2-5 LU/ha, finishing on forage and supplemental concentrates, high growth rates, slaughter at 12-18 months depending on seasonality and time of birth.
Mixed semi-intensive/extensive grazing and intensive finishing systems; indigenous and improved indigenous or temperate cross beef breeds;	Seasonal grazing +/- conserved forage based systems, improved natural pastures to extensive rangelands or uplands (ranch or communal grazing), often with marked seasonality, stocking densities <2 LU/ha, some supplementary concentrates used in breeding herd, intensive mixed forage and concentrate feeding for fattening (may be grazing or feed-lot based, feeding period 3-5 months), usually with intermediate store or over-wintering/dry season phase, use of variety of conventional and unconventional locally available concentrates, moderate to low overall growth rates, slaughter 18-24 months (semi-intensive) and 2-4 years (extensive).
Small-holder stall-fed fattening systems; indigenous and some improved indigenous and crosses	Stall-fed stored crop residues and crop processing by-product concentrates, feeding periods 3-12 months, slaughter 2-5 years.

Poultry and pig production systems range from the most intensive large scale systems based entirely on compound concentrate feeding to low-intensity traditional subsistence production systems. The latter utilise concentrates mainly in the forms of crop or human food processing by-products and wastes, though because of their wide distribution throughout the world consume appreciable quantities of these feeds.

Cattle production systems similarly range in intensity from the most intensive feed-lot dairying and cereal-fed beef systems to small-holder subsistence dairy production. Ruminants generally utilise a wider range of conventional and non-conventional feeds than monogastric animals, even in the most intensive systems. In total, small-holder dairy and beef production systems probably do not account for a large proportion of concentrate feeding, though they may be the main users in some localities.

Milk and meat production systems based entirely on forages are not included in the classification in Table 1. These include grassland based milk production such as in New Zealand, and the many extensive milk and meat production systems with various species of livestock. These systems make virtually no use of concentrate feeds and opportunities for their intensification by this means in the future are limited.

Occurrence of Concentrate Feeding Systems

The mix of livestock production and feeding systems that utilise concentrate feeds varies across the different farming systems and geographical regions of the world. Table 2 illustrates the occurrence of concentrate feeding systems within the farming systems recently classified by FAO. The latter distinguish livestock production in landless, grassland-based and rain-fed or irrigated mixed farming systems in (a) the temperate and tropical highlands, (b) the humid and sub-humid tropics and sub-tropics, and (c) the arid and semi-arid tropics and sub-tropics.

Table 2. Occurrence of main livestock production systems utilising concentrate feeds within farming systems

Farming Systems	*Livestock Production systems using Concentrates*
Rainfed mixed crop-livestock	
Temperate and tropical highlands (MRT)	*Cattle milk:* - intensive grazing/forage-based (eg N America, WE, EE) - small-holder (eg E African highlands, India, Nepal, S and C America); *Beef production:* - intensive grazing/forage and concentrate-based systems (eg N America, WE, EE) *Poultry and pig production:* - semi-intensive and traditional small-holder
Humid/sub-humid tropics and sub-tropics (MRH)	*Cattle milk:* - intensive grazing/forage-based (eg tropical USA and Australia) - small-holder (eg lowland SS Africa, Asia); *Beef production:* - semi-intensive mixed farming dual purpose systems (eg C and S America) - small-holder stall-feeding (eg Malawi, N Nigeria) *Poultry and pig production:* - semi-intensive and traditional small-holder
Arid/semi-arid tropics and sub-tropics (MRA)	*Cattle milk:* - semi-intensive and small-holder (eg SS Africa, C America, Brazil, India, N Africa, W Asia) *Beef production:* - some dual purpose semi-intensive and small-holder (eg Mexico)

	Poultry and pig production: - semi-intensive and traditional small-holder
Irrigated mixed crop-livestock	
Temperate and tropical highlands (MIT)	*Cattle milk:* - intensive grazing or zero-grazing (eg Netherlands, Italy, Japan, EE, China) - semi-intensive small-holder (eg China, India, Yemen) *Beef production:* - some intensive grazing/forage based (eg Italy, Japan, EE) *Poultry and pig production:* - semi-intensive and traditional small-holder
Humid/sub-humid tropics and sub-tropics (MIH)	*Cattle milk:* - semi-intensive mixed grazing/ stall-feeding (eg India, S and E Asia), - small-holder (widespread Asia, including buffaloes) *Beef production:* - some dual purpose (Asia) *Poultry and pig production:* - semi-intensive and traditional small-holder
Arid/semi-arid tropics and sub-tropics (MIA)	*Cattle milk:* - intensive zero-grazing (eg Israel, W Asia, N Africa, California/ USA); - semi-intensive fodder crop-based (eg W Asia, N Africa, S Asia); - small-holder (W Asia and S Asia) *Beef production:* - some dual purpose (C America, N Africa)

	Poultry and pig production: - semi-intensive and traditional small-holder
Grassland based livestock	
Temperate and tropical highlands (LGT)	*Cattle milk:* - semi-intensive grazing with very little concentrate feeding (eg New Zealand, Australia, Kenya highlands, S and C America highlands); *Beef production:* - some dual purpose semi-intensive grazing (N Zealand, Australia, S America highlands) *Poultry and pig production:* - semi-intensive and traditional small-holder
Humid/sub-humid tropics and sub-tropics (LGH)	*Cattle milk:* - semi-intensive grazing (eg S and C America lowlands, Caribbean) *Beef production:* - semi-intensive grazing (eg S America lowlands, Argentina, Uruguay, N Australia, SS Africa) - some dual purpose semi-intensive grazing *Poultry and pig production:* - semi-intensive and traditional small-holder
Arid/semi-arid tropics and sub-tropics (LGA)	*Milk production:* - semi-intensive grazing (Australia, S Africa, S America) *Beef production:* - mixed extensive grazing +/- intensive finishing systems (eg Zimbabwe, S Africa, Australia, W USA) *Poultry and pig production:* - semi-intensive and traditional small-holder

Landless livestock	
Landless monogastrics (LLM)	Poultry eggs, poultry meat, pig meat: - intensive and semi-intensive systems (all regions, especially developed countries, cereal producing areas, peri-urban areas)
Landless ruminants (LLR)	*Cattle milk:* - none (all partly forage based) *Beef production:* - cereal-based fattening systems.

Intensive pig and poultry production systems are classified as landless monogastric systems. These systems occur in any part of the world wherever market demand, feed supplies and the availability of support services allow. They are usually associated with centres of human population and higher levels of income. Semi-intensive and traditional small-holder pig and poultry systems occur widely in mixed farming systems across all regions of the world.

Types of ruminant production systems are more determined by agro-climatic conditions. Cattle milk production systems based on intensive grazing or cut forages occur especially in temperate Europe and N America, but also in parts of Oceania. These systems make intensive use of productive sown pastures and forages, with stocking densities typically in the range of 2-5 LU/ha. They use large amounts of concentrate feeds. Intensive zero-grazing systems based on irrigation occur mainly in semi-arid and arid areas, utilising irrigated fodders and concentrate feeds. Semi-intensive and small-holder cattle (and buffalo) milk production systems occur quite widely within both rainfed and irrigated mixed farming systems. They can occur in all agro-climatic zones but are less common in the arid and semi-arid zones (unless partially irrigated) because of problems of feed and water supply.

For beef production, intensive grazing and cereal feeding systems also occur mainly in the temperate zones. Mixed semi-intensive and extensive systems occur mainly in semi-arid zones, with some in sub-humid zones.Thus, it is evident that a wide variety of concentrate feeding systems occurs within each of the broad classes of farming systems outlined by Sere. Aggregate feed demand and utilisation within broad farming systems and regional zones is thus difficult to estimate.

Utilisation of Concentrate Feeds

According to the most recent global review, concentrate feeds constituted about a quarter of all feeds fed to livestock in the mid 1980s. This proportion was growing at the rate of about one percentage point every five years to that time. In developed countries concentrates comprised almost 40% of total feeds but in developing countries only 12%. Cereals constituted half to three-quarters of this concentrate, with most of the remainder provided by cereal milling residues and oilmeals. Pulses provided about 2% and roots and tubers 4.3% of total concentrate feeds dry matter in 1986.

Other concentrates are provided by ingredients such as molasses, dried grass and other forage products, milk and other animal derived products, and agro-industrial by-products such as citrus pulp, sugar beet pulp and brewing wastes. Estimates of the total quantities of these components are not readily available. Estimates based on FAO data in the early 1980s suggest that they may provide a further 12-15% of the total concentrates as defined above. In the EU in 1986/87 these feeds provided 12% of overall concentrates in metabolisable energy terms. A global overall figure for the contribution of these feeds to total concentrates has been assumed in following analyses to be 12%.

Table 3 summarises FAO Agrostat data on the global total utilisation for feeds of cereals, brans, oilseeds, oil cakes and meals, and roots and tubers over the period 1965/67 to 1990/92. Annual total consumption of these concentrate feeds amounted to 982 million mt averaged over 1990/92.

Table 3. Global total utilisation of concentrate feed resources 1965/67 to 1990/92

	Global total utilisation for feeds (1) ('000 mt per year)			
Commodity	1965/67	1975/77	1985/87	1990/92
Cereals	354,808	496,343	629,802	600,516
Brans	72,359	87,155	113,222	118,946
Oilseeds	3,291	4,129	9,666	13,463
Oilmeals and cakes	41,301	66,201	103,631	119,263
Sub-total	471,759	653,830	856,322	852,118
Roots and tubers	131,328	126,606	133,243	129,860
Total	603,087	780,436	989,565	982,049

Cereals constitute the major proportion of concentrate feeds, amounting to over 60% of the total including roots and tubers. Brans, oilseeds, oilcakes and roots and tubers constituted 12%, 1.4%, 12% and 13% respectively of total concentrates in 1990/92. The contributions of cereals and brans to the total have remained constant over the recent 25 years, while oilseeds and oilcakes have increased markedly and roots and tubers declined over the period.

Global total consumption of all feeds increased consistently over the period 1965/67 to 1985/87, by 29% and 26% respectively in the two decades to 1985/87. Latterly, since 1985/87, total consumption has stabilised mainly due to recessions in OECD countries and restructuring in former centrally planned economies. Consumption of oilseeds and meals has continued to grow particularly strongly throughout the period, including since 1985/87. Table 4 shows the annual percentage growth rates in the utilisation of different types of feeds.

Table 4. Average annual growth rates in global total utilisation of concentrate feed resources 1965/67 to 1990/92

	Average annual growth rates in utilisation of feeds (% per year)		
Commodity	*1965/67-1975/77*	*1975/77-1985/87*	*1985/87-1990/92*
Cereals	3.41	2.41	-0.48
Brans 1.88	2.65	0.49	
Oilseeds	2.29	8.88	3.37
Oilmeals and cakes	4.83	4.58	2.85
Roots and tubers	-0.37	0.51	-0.51
Total	2.61	2.40	-0.15

Consumption has slowed or declined since 1985/87 in North America (+3%), Western Europe (-10%) and E Europe and the states of the former USSR (CIS) (-21%) but has continued strongly in East Asia (+18%) and in most other developing regions. Over the whole period from 1975/77, consumption has more than doubled in W Asia and N Africa (WANA) and SE Asia, and increased by 61% in S Asia respectively, though from relatively low bases. In contrast, increases in Sub-Saharan Africa (SSA) have been only 33% and in Central and Southern America (CSA) 43%. Of most significance globally is the rate of increase on an already large base in E Asia.

Trends in annual average growth rates of consumption of total concentrates in different world regions are illustrated in Table 5. Growth rates have generally declined over recent periods in developed regions.

Table 5. Average annual growth rates in regional utilisation of total concentrate feed resources 1965/67 to 1990/92

	Average annual growth rates in utilisation of feeds (% per year)		
Region	*1965/67-1975/77*	*1975/77-1985/87*	*1985/87-1990/92*
Developed regions			
North America	0.39	2.59	0.61
Western Europe	2.34	1.27	-2.03
E Europe and CIS	5.01	0.82	-4.55
Oceania	2.07	4.36	7.44
Developing regions			
C and S America	3.62	3.44	0.46
W Asia and N Africa	4.30	6.52	2.61
Sub-Saharan Africa	2.64	2.19	1.40
S Asia	2.59	2.72	4.25
SE Asia	4.30	6.73	6.29
E Asia	1.25	4.42	3.40
World	2.61	2.40	-0.15

Growth has been more consistent and more buoyant in developing regions, except for Sub-Saharan Africa and Central and Southern America in recent years. Growth has been particularly strong in all Asia regions. Three regions, N America, EE+CIS and E Asia now each consume over 20% of world concentrate feeds, and W Europe a further 16%, amounting to 80% of the world total concentrates. Poorer developing regions still have very low levels of consumption such as in SSA (1%), S Asia (3%) and SE Asia (3%), despite high populations.

Developed regions are still major users of cereals, oilseeds and meals, with over 60% of global consumption of each, while developing regions make proportionally

greater use of brans and roots and tubers. Brans are particularly important in Asian regions, where rice bran is a major feed resource. Roots and tubers consumption is highest in E Asia and in E Europe and CIS.

Changes in regional patterns of consumption follow the annual growth rate trends outlined above. Most notably, the proportions of world consumption utilised in WE and EE+CIS have declined, with similar declines across all feed types. Meanwhile, the share of world consumption in E Asia has increased from 15% to 22%. Developing regions in general have increased their share of feed utilisation, a trend which will continue as growth in demand for livestock products continues in these regions.

Composition of concentrates fed in world regions: In developed regions, cereals consistently provide the major proportion of concentrate feeds, varying in 1990-92 from 59% in WE to over 80% in NAM and Oceania (PAC). Cereals are the main concentrates also in CSA and WANA. In other developing regions, brans are relatively more important, providing 20-50% of total feeds. Oilseeds and meals are significant resources in all regions, but especially in S Asia (SASIA) where they constituted a total of 41% of available feeds in 1990/92. Roots and tubers are most significant in E Asia and CSA, though still also important in EE+CIS (19%) and WE (10%).

The most significant changes in the composition of concentrates have included increases in the proportions of oilmeals used across all regions of the world, but particularly in WE, CSA, S Asia and SE Asia. In NAM and WE there has been a shift in concentrate composition from cereals to oilseeds and meals, with relatively little change in other components. Cereals have increased and brans generally decreased as proportions of total concentrates in the more rapidly developing regions of

CSA, SE Asia and S Asia, in contrast to the patterns in SSA where cereal consumption has declined further from an already low base in 1975/77. The relative importance of roots and tubers feeds has declined consistently across all regions. Concentrate feed quality has thus generally increased as total consumption has increased.

Utilisation of Specific Concentrate Feed Commodities

World total amounts of the main specific feed commodities utilised are shown in Tables 6, 7 and 8 for cereals, oilmeals and roots and tubers respectively. Amongst cereals, maize is the most important feed, comprising over half of total cereals. Wheat and barley each comprise some 20% of total cereals, while sorghum, oats and millet together constitute less than 10%. Quantities of maize and wheat consumed as feed have increased progressively over the period 1975/77 to 1990/92. Barley, oats and sorghum utilisation has recently declined.

Table 6. World total annual consumption of different cereals for concentrate feeds in 1975/77 to 1990/92

	Annual utilisation of different cereals ('000 mt)					
Year	*Maize*	*Wheat*	*Barley*	*Oats*	*Sorghum*	*Millet*
1975/77	217976	71953	109405	37791	31937	1695
1985/87	308545	106651	129944	37437	38115	2752
1990/92	330796	118142	120805	27223	32426	2385

Other cereals used in concentrate feeds, not listed in Table 6, include rye and small amounts of buckwheat, triticale, fonio and quinoa. In addition, small amounts of rice are fed in Asian regions, amounting to some 5-6 mill. mt/year. These form an insignificant proportion of cereal use. Amongst oilmeals and cakes, soyabean dominates, comprising over 60% of total oilmeals consumed. Cotton

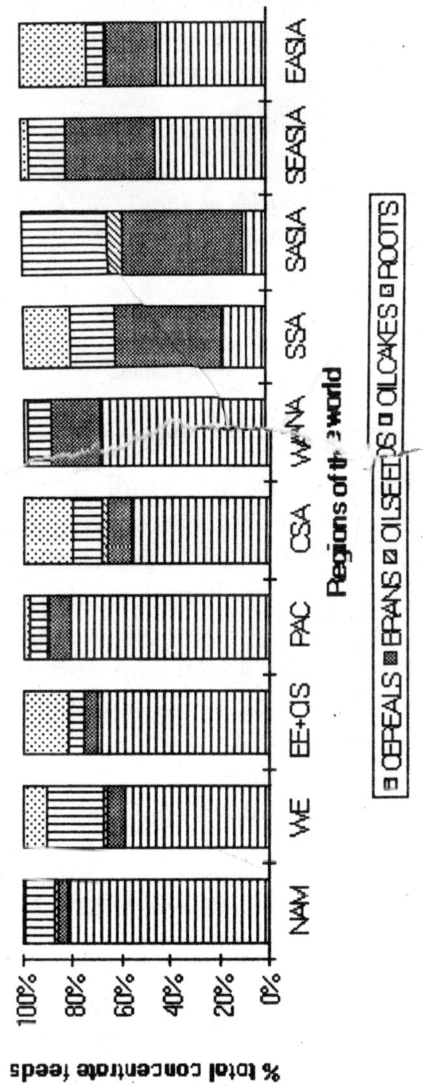

Figure 1. Composition of concentrate feeds utilized in regions of the world 1990/92

seed, rape and sunflower oilmeals each comprise about 10% of total consumption. Groundnut and palm kernel meals are of lesser importance globally. In addition to the tabulated commodities, smaller amounts of meals and cakes of copra, hempseed, mustard, linseed and safflower are fed.

Table 7. World total annual consumption of oilmeal commodities for concentrate feeds in 1975/77 to 1990/92

	Annual utilisation of different oilmeals ('000 mt)					
Year	*Soya bean*	*Ground nut*	*Cotton*	*Rape*	*Sun-flower*	*Palm kernel*
1975/77	41911	3693	7201	3485	4324	603
1985/87	64995	4221	8962	8362	7610	1368
1990/92	71842	4824	10297	11101	8525	1808

Fishmeal provides a further 5-6 mill. mt/year. Utilisation of all major oilmeals has increased progressively over the period 1975/77 to 1990/92, but particularly of soyabean, rape, sunflower and palm kernel meals. Oilseeds constitute a relatively small proportion of total concentrates. Amongst the seeds, soyabean and cotton seed are the main commodities making up 41 and 29% respectively of seeds fed. The most important roots and tubers are potatoes and sweet potatoes, these each comprising about 40% of total roots and tubers consumption.

Table 8. World total annual consumption of different roots and tubers commodities for concentrate feeds 1975/77 to 1990/92

	Annual utilisation of different roots and tubers ('000 mt)				
Year	*Cassava*	*Dried Cassava*	*Potatoes*	*Sweet potatoes*	*Yams*
1975/77	17095	3288	80179	25595	305
1985/87	17287	7133	66720	41784	369
1990/92	17012	8668	52511	46218	381

On a dry matter basis, cassava, dried cassava, potatoes and sweet potatoes are each of similar importance, each comprising 20-25% of total consumption. Yams and other roots and tubers contribute less than 1% of the total. Feed consumption of potatoes has declined over 1975/77 to 1990/92. Sweet potato consumption has almost doubled in the same period, mainly due to increased production in China.

Maize is of particular importance in N America and CSA, but also in developing regions of SSA and Asia. It is relatively unimportant in W Europe and EE+CIS and Oceania (5%). In these regions, wheat and barley are the main cereals, each comprising about 30-40% of the total. Barley is particularly significant in WANA where it comprises over 50% of total cereal feeds. Sorghum is of lesser importance overall, but provides over 20% of cereal feeds in CSA and S Asia.

Amongst the oilmeals, soyabean is almost universally important except in Oceania, SSA and S Asia. In these areas, and in WANA, cotton seed meal is particularly important, providing 30-40% of total oilmeal feeds. Groundnut is locally significant in SSA, S Asia and SE Asia. Rape seed meal is notably important in S Asia and utilisation has been growing in WE and Oceania. Sunflower contributes over 10% of oilmeal feeds in WE, EE+CIS, WANA and SSA.

Utilisation of roots and tubers feeds differs strongly between regions. Cassava is of primary importance in CSA, SSA and SE Asia. Dried cassava is prepared especially for WE and WANA markets. Potatoes are particularly important in EE+CIS. Sweet potatoes are used almost exclusively in E Asia and to a lesser extent in SE Asia.

Use of Concentrate Feeds by Livestock species

Concentrate feeds generally form a higher proportion of diets of monogastric than of ruminant livestock. Figure 2 illustrates global average diet compositions of different species reported by FAO. Concentrate feeds make up over 80% and 90% respectively of the diets of pigs and poultry, though only 10% and 2% of the diets of the overall cattle/buffalo and sheep/goat populations respectively. Pig and poultry diets include, on average, some 50-60% cereals, 10-20% oilmeals and 15-25% 'other concentrates'.

Diet compositions vary between species and regions of the world. Pig and poultry diets in developed countries typically include 40-60% and 55-65% cereals respectively, and less than 15% milling by-products. In developing countries, diets generally include less cereals and more by-products. Typical formulations of compound feeds in India, for example, include 30-45% cereal brans.

Ruminant concentrate feeds include higher proportions of 'other concentrates'. For dairy cattle in UK, for example, diets typically include 40-50% cereal milling by-products, 25% oilmeals and 20% of other ingredients such as molasses and sugarbeet pulp, though still including some 10-15% cereals. Some feeding systems for fattening beef cattle use high proportions of coarse grains as straight concentrates but in most cases a wide variety of milling and agro-industrial by-products may be incorporated in finishing diets. In developing countries, the proportion of cereals in concentrates fed to ruminants is very low.

However, because of their large numbers and size, ruminant livestock actually consume roughly similar total quantities of concentrate feeds as pigs or poultry. In metabolisable energy terms, cattle, small ruminants, pigs

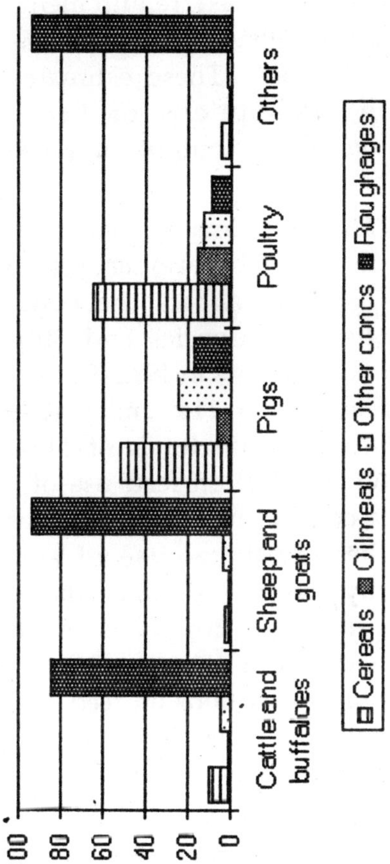

Figure 2. Aggregate world composition of diets for different species of livestock (after Wheeler et al 1981; in metabolisable energy equivalents)

and poultry consumed averages of 35%, 3%, 33% and 29% respectively of total concentrate feeds in 1981.

Concentrate Feed Requirements in Farming Systems

Estimates of concentrate feed requirements have been made in this study for the farming systems and world regions classified by Sere. These estimates have been based either on assumptions of per capita feed requirements, or on product outputs and feed conversion efficiencies.

Table 9 presents the estimated total concentrate feed requirements for the different farming systems and world regions, while Table 10 shows the consumption of concentrates by livestóck species and outputs. Global total concentrate requirements in 1992/93 are estimated at 996 mill. mt dry matter, or 1145 mill. mt fresh weight. This may be compared with totals of 1142 million mt derived from FAO statistics on feed use of 982 mill. mt cereals, brans, oilseeds and meals and roots and tubers in 1990/92, plus the additional use of some 26 mill. However, it may be noted that the relative or proportional requirements for feeds by the different systems, regions and species are not particularly sensitive to minor changes in assumptions on feed efficiencies and intakes, being mainly dependant on estimates of livestock populations and total outputs of products within systems and regions.

OECD developed countries consume over a third (37%) of world total concentrates, with Asia and EE+CIS taking about a quarter each. Amongst the production systems, Landless Monogastric (LLM) systems consume the most concentrates. Mixed rainfed temperate systems (MRT), mainly consisting of dairy cattle in N America, Western Europe and EE+CIS, consume about a quarter of the total. Landless Ruminant (LLR) systems consume

Table 9. Estimated annual utilization of concentrate feeds for all livestock production by world regions and production systems in 1992/93 ('000 mt dry matter)

	Concentrate feed use for all livestock production by regions of the world ('000 mt dry matter/year)							
Production system	*SSA*	*WANA*	*ASIA*	*CSA*	*OECD*	*EE+CIS*	*OTH DEVP*	*Overall*
LGT	1,398	798	492	1,929	3,687	0	0	8,304
LGH	1,070	0	271	1,985	887	0	1,110	5,323
LGA	1,782	567	0	392	2,391	0	1,029	6,162
MRT	2,244	0	1,039	1,505	132,222	118,728	0	255,738
MRH	4,733	0	15,584	19,827	6,459	0	0	46,603
MRA	2,431	9,968	20,649	2,682	198	0	0	35,928
MIT	0	0	30,122	4,208	29,489	9,629	0	73,448
MIH	0	0	81,014	389	1,929	0	0	83,332
MIA	51	12,066	33,558	2,493	6,275	0	457	54,900
LLM	2,291	6,616	89,015	31,324	160,302	38,824	2,214	330,586
LLR	0	1,165	0	0	19,794	74,517	0	95,476
Overall	15,999	31,180	271,743	66,735	363,634	241,699	4,810	995,800

Table 10. Estimated annual utilization of concentrate feeds for different livestock species and products by world regions and production systems in 1992/93 ('000 mt dry matter)

	Concentrate feed use by regions of the world ('000 mt dry matter/year)							
Species/ product	*SSA*	*WANA*	*ASIA*	*CSA*	*OECD*	*EE+CIS*	*OTH DEVP*	*Overall*
Beef, veal and buffalo meat	1,423	450	8,196	4,401	40,586	77,297	251	132,604
Small ruminant meat	312	1,908	1,771	256	2,839	1,381	78	8,545
Cattle milk	5,035	16,703	47,277	12,709	85,486	76,524	829	244,564
Pig meat	2,114	21	128,617	13,416	135,744	53,885	639	334,436
Poultry meat	3,612	6,535	30,789	21,961	62,743	14,139	1,954	141,733
Poultry eggs	3,503	5,563	55,092	13,991	36,235	18,473	1,060	133,917
Overall	15,999	31,180	271,743	66,735	363,634	241,699	4,810	995,800

some 10 %, mainly for beef production, while the remainder (32%) is distributed evenly amongst the remaining mixed farming systems.

Of the specific combinations of systems and regions, LLM and MRT systems in OECD countries, MRT and LLR systems in EE+CIS, and LLM systems in Asia each consume between 8-15% of total concentrate feeds.

Amongst the species and products, pig meat production is estimated to consume the largest proportion of feeds (34%), especially in OECD countries and in Asia. Cattle milk production consumes almost a quarter of the total, mainly in OECD and EE+CIS, while poultry consumes a further quarter roughly evenly split between meat and egg production. Beef production is estimated to consume about 13% of the total.

Overall, the consumption of concentrate feeds is split roughly 60% to monogastric livestock and 40% to ruminants. Amongst the regions using most concentrates, monogastrics' share of feeds appears highest in Asia (74%) and lowest in EE+CIS (33%) where ruminants consume the largest proportion (66%).

6

Crop-Livestock Interactions

The nature of the associations between crops and livestock is an issue that needs to be considered if crop, soil and livestock models suited to ILRI's objectives in adopting a systems modelling approach are to be identified. The interfaces between livestock and other system components define, implicitly, the data that are required to characterise resource flows and other interactions. For example, if the outputs of a livestock model are to be used to drive a soil model, the inputs that the latter requires will prescribe what these outputs should be; at least, this will be the case if structural changes to the soil model are not envisaged. This principle applies, irrespective of the level of aggregation of the models being considered. Furthermore, an appreciation of it helps to establish, from a common sense perspective, that the components of a systems model need to be implemented at similar levels of detail.

Figure 1 represents, at the most aggregated and general level, the components of an agricultural system and their linkages. The extent to which each of these components would need to be modelled in order to simulate the behaviour of a given system may vary considerably depending on the nature of that system. Figure 1 also illustrates several aspects of agricultural systems that are relevant to their representation as models and, therefore, on which the recommendations of this review will hinge:

- *There is scope for considerable structural differences amongst systems* even when they are represented at this, most simple, level. Livestock may or may not be present. The contribution of off-farm resources such as range or forest to the farm's organic resource pool may be highly significant or minimal. Models implemented at a more disaggregated level than Figure 1 can only, therefore, hope to be as reliable as the system characterisations that underpin their design or operation;
- *The farming system is, essentially, cyclic* at its core (organic resources - livestock - land - crops). Therefore, management decisions in one component may produce an impact in all others. By extension, these decisions can also feedback on the component being manipulated in the first place as a result of the changes induced in other components. It is this feature of agricultural systems that makes modelling a particularly powerful tool in their analysis. Controlling variation in experimentation across multiple system components to an extent that will allow the identification of these subtle but often highly significant sources of variation is rarely likely to be feasible;
- *Resource competition is a significant issue at most decision points in the system.* Decisions on the optimum allocation of scarce resources are often made, by farmers and planners alike, with an incomplete knowledge of their impacts. Modelling approaches should seek to fill in at least some of the gaps.

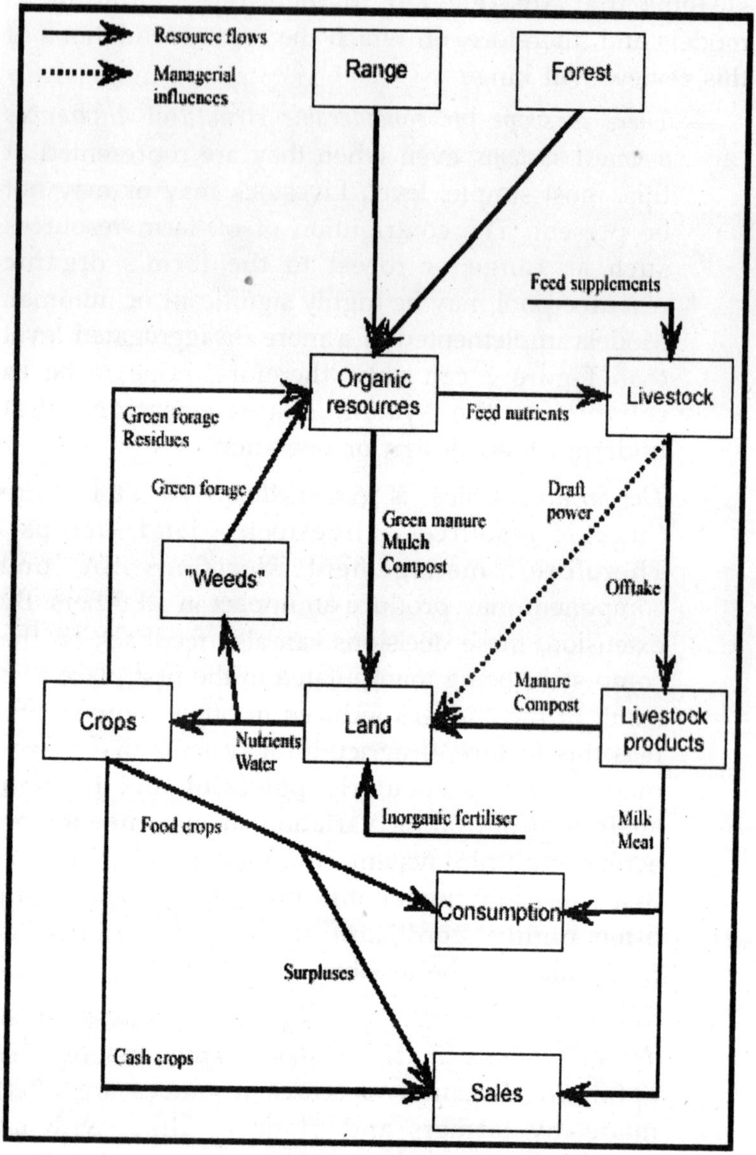

Figure 1. A general, schematic representation of the components of agricultural systems and their interactions.

Key Interfaces in Crop-Livestock Systems

The following key interfaces will need to be considered if livestock models, integrated with crop or soil models, are to be used successfully to simulate the operation of crop - livestock systems with these objectives in mind:

— Organic resources - Livestock

— Livestock - Land

— Livestock product utilisation

At each interface there may be a number of processes operating and an integrated model should be able to account both for the individual effects of these processes and their interactions.

Organic Resources - Livestock

The interface between organic resources - of both on-farm and off-farm origin - and livestock largely represents the supply of nutrients and energy in feed. Some organic matter may also be used as bedding material for animals. However, this is generally associated with the effective trapping of voided nutrients in faeces and urine and is, therefore, more properly associated with the livestock - land interface.

The management of nutritional inputs at the organic resources - livestock interface may be achieved in two distinct ways which need to be treated differently when modelled:

— indirect management in a grazing or browsing situation;

— direct management of feed offers to stall-fed livestock.

The management of grazing usually leaves animals to exercise a considerable degree of self-determination in what they actually consume. The length of access time and the general geographical area may be controlled by the herder but, unless grazing a singlespecies, planted pasture, it is likely that the animal will have the freedom to choose a varied diet. The prediction of quantitative intake and the mix of species consumed is a matter that has occupied grazing ecologists for a number of years. This issue must be addressed by models wherever grazing is a major element of the system under study.

Stall-feeding is common in mixed farming systems as it allows farmers to exert a greater degree of control over the valuable manure outputs of their animals and also reduces the possibility of damage to crops that may be caused by free-ranging livestock. This managerial control makes the simulation, in modelling studies, of the quantities and types of feeds used easier than it is in the grazing situation. However, pure stall-feeding is probably relatively rare.

The extra labour required for feed collection is not always available. Furthermore, where communal grazing *is* available, not to use it may represent a waste of a potential resource that a farmer will not countenance. Thus, livestock in "stalled" systems will often be grazed as well - usually at a particular time of the year when seasonal factors make this the most desirable management option.

Livestock - Land

The livestock - land interface includes two, quite distinct forms of interaction:

— the production of manure and compost;

— the provision of Draught animal power;

both of which may exert a considerable impact on the productivity of the cropping system.

The pivotal role of livestock in the cycling of nutrients to crops, wherever the two are associated, has gained considerable ground in recent years as a topic of interest to both the research and the development communities. It is not under dispute that livestock can exert considerable, beneficial effects on system productivity through the manure (and compost) - soil - crop pathway, improving both chemical and physical properties of the soil for plant growth. However, few attempts have been made to derive an integrated, mechanistic understanding of the ways in which dynamic processes in livestock become interlaced with dynamic processes in soils.

Many current soil models rely on superficial characterisations of manure quality. In situations where the good husbandry of manure to preserve its quality is unknown, this may be justifiable as a large proportion of the nutrients that leave the livestock component of the system by this pathway never find their way to the soil. However, given the current interest in the development of improved composting and manure management practises to maintain quality, the considerable effects of livestock management on that quality are likely to assume greater significance.

Of the crop - livestock interactions that are pertinent to a consideration of the potential for modelling mixed farming systems, the provision of Draught animal power must be viewed in a separate light. Other interactions represent resource flows from one system component to another. Draught animals are used, essentially, as tools in the management of another system component; mainly, in this context, the soil through tillage operations although their role in support of crop processing and marketing should not be underestimated in some situations.

Farmers perceptions of Draught animals in mixed farming systems are often somewhat ambiguous. Access to them is viewed as essential at times of the year when land preparation operations must be carried out. At other times their feeding and general care may be viewed as a, largely unproductive, chore. For this reason, Draught animals may often be managed on a basis of adequacy with little scope for major increases in productivity through improved feeding or management. Nevertheless, Draught animals are consumers of and, therefore, competitors for farm resources. Where they are significant constituents of a mixed farming system, their impact on the management of that system cannot be ignored in attempts to simulate its operation.

Livestock Product Utilisation

The availability of organic resources and constraints on their utilisation generate a supply side driving force in mixed farming systems. On the demand side, this role is generally played by an array of economic considerations related to on-farm utilisation and sales of products from both crop and livestock enterprises. These define the overarching objectives that farmers pursue in their farming activities and include the provision of a year-round food supply for the family and the generation of wealth and cash flow to cover other expenses such as schooling and medical costs.

A key question here is whether it is desirable to address these issues *within* the systems modelling framework or whether they should be regarded as external - that is, considered only during the processes of quantifying input variables and evaluating output scenarios. A number of economic models have been constructed to examine the behaviour of farmers in pursuit of household food security and income generation. These are generally, and often quite justifiably given their objectives, based on simplified coefficient-based treatments of the biological processes occurring in production systems. As this approach precludes an evaluation of the consequences of interactions between processes in different components of mixed farming systems, such models are unlikely to be well-suited to a detailed examination of the impacts of component interventions at the whole system level. This, of course, opens the way for an inverse criticism.

The practical utility of deterministic, biological models may be limited by their inability to place their outputs in the context of the benefit that might actually be conferred upon the farmer who adopts the strategies being tested. A "third way" that has gained some ground in recent years is the development of bio-economic models which pursue the sound, in methodological terms, strategy of attempting to model all these aspects of the system at a similar level of detail.

Some bio-economic models have already reached quite high levels of detail and sophistication and, undoubtedly, may make a contribution in the future. However, the suspicion must always remains that the complexity and degree of unpredictability in farmers' behaviour may not always be adequately assessed, undermining the reliability of the predictions of such models and, therefore, the acceptability of any recommendations based upon these.

This is a debate that has some way to run. However, it is an issue that has some bearing on how the modelling activities that develop from ILRI's initiative in this area will ultimately exert an impact. What should be a *sine qua non* is that future, biological modelling activities are integrated with an appreciation of farmers' objectives, at all levels, in the systems that are being targeted.

Possibilities of Crop- Livestock Interactions

On a global scale, there is a considerable range of agricultural systems in which tangible associations between crops and livestock may be observed. The precise nature of these associations also vary considerably. Similarly, the literature reports a substantial variety of crop, soil, livestock and farming systems models which describe aspects of these associations to address different objectives with varying degrees of success.

Associations between Pastoralists

Figure 2 illustrates the main associations between the livestock kept by pastoralists and the cropped land farmed by settled agriculturalists. This is generally a transient association that occurs at times of the year when crop residues or other feed resources are available on farms and allows the transfer of nutrients in animal manure and urine to the cropped land. The incidence of tsetse fly may also affect the extent of penetration by transhumant pastoralists of the more humid areas in which cropping is concentrated.

The characterisation of resource transfers between livestock and crops are likely to be simpler, from a modelling perspective, than would be the case with a more integrated crop-livestock system. Animals, are usually kept directly on the cropped land in order to use residues so that the more complex considerations associated with feed collection and manure/compost handling that arise in stall feeding situations do not generally occur. However, the variable temporal and spatial aspects of the association may mean that the implications of changes in it are not always immediately apparent and may be more far-reaching in terms of productive potential at, for example, a regional level.

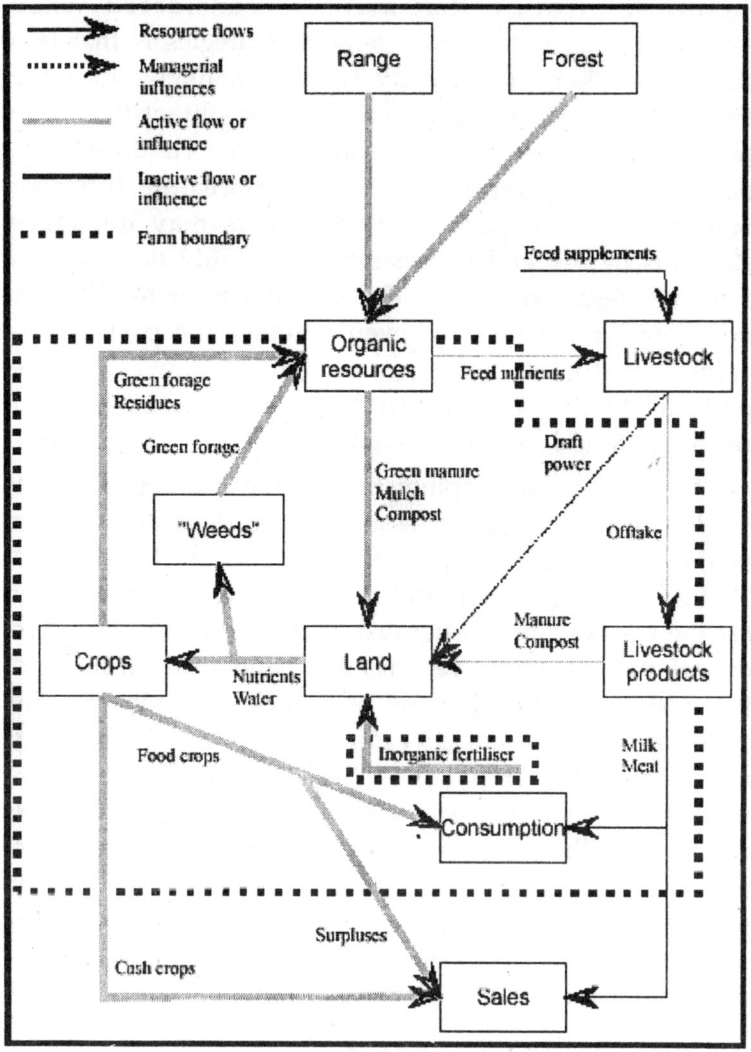

Figure 2. Schematic representation of the interactions between settled agriculture and livestock belonging to pastoralists.

The nature of the interaction is determined largely by the agreement reached between the pastoralist and the settled farmer to exchange excreta for feed resources. A significant feature that must be considered when examining alternative management strategies is that two decision-makers are involved. Each is likely, to some extent at least, to have conflicting objectives. The resolution of these conflicts can, in some circumstances, become quite complicated. Furthermore, the different objectives of the participant managers may introduce elements of risk and uncertainty that would not occur in a more integrated system. The settled farmer has little or no control over when or even whether the pastoralist's animals will come onto his or her land. Conversely, the pastoralist will not be able to determine the availability of residues in advance in a given area.

These issues do impinge on how best to approach the modelling of this type of association. Intervention testing in the agricultural component will generally be directed at the cropping system and may not adequately account for the needs of the pastoralist. If the modelling objective to seek management strategies that benefit the farmer and the pastoralist, then the use of models that consider the circumstances and objectives of both will be needed. This means that the impacts of the biological and socio-economic constraints on all participants will need to be addressed. Furthermore, some a formalised multiple objective analysis may be preferred over simple biological models of the cropping system which characterise the associated pastoral system only superficially amongst the data that are used as inputs for simulations.

Key Elements of Livestock Models

Attempts to simulate the behaviour of livestock populations must consider a number of key sets of processes. Some of the studies discussed have been conducted at a level of detail greater than that which would generally be found in the components of a livestock systems management model. A consideration of these will, however, highlight some of the generalisations implicit in management models that may affect the reliability of predictions based upon their use. Some general principles which should be heeded when attempting to integrate livestock into systems modelling activities will be established.

Resource Utilisation

Diet selection and intake under stall feeding conditions

Under stall feeding conditions, quantities of feeds offered to livestock are generally controlled by the farmer and may, therefore, be specified directly as input variables for simulation models. In situations where feeding rates are low in relation to the intake limits of the animals being fed, refusal rates are also likely to be low. However, in some circumstances — and in particular where livestock are tightly integrated with crops — farmers may deliberately offer feed at levels above those that they know their animals will readily consume. This can allow the selection of more nutritious fractions of the offered diet and for an "auto-composting" process to take place as feed refusals become blended with faeces and urine.

Tanner *et al* have described the effects of increasing offer rates to goats fed indigenous grasses in Indonesia. Increasing forage offer rates from 25 to 75 g/kg bodyweight/day resulted in an increase in refusal rates of almost 500%. An effect of this magnitude may be expected to complicate the estimation of nutrient intakes in stall feeding situations for modelling purposes as the total quantity of feed consumed can no longer be assumed to be the same as the amount offered. Furthermore, diet selection may result in improved nutritional value of the consumed feed in comparison with the composition of representative samples of the feed offered.

The best, simple approach for use in management models that is available at present would seem to be one that would introduce a bias in favour of the more nutritious components of the diet as the overall feeding rate is increased. This would need to be implemented with, at least, a level of species-dependence to account for differences in the selective abilities of small and large ruminants.

Diet selection and intake in grazing animals

Attempts to model diet selection and intake in grazing animals have been much more widespread; perhaps because these are issues that are pertinent to the production systems found in industrialised countries. Much of the difficulty in reliably predicting the consumption and subsequent utilisation of grazed forages by animals arises from the interplay between plant and animal factors and their combined effects on the quantity and quality of herbage consumed.

In most attempts, there is, in the interests of basic feasibility, some compromising of consistency in the level of detail of the components of these models and mixtures of empirical and mechanistic treatments.

The model of Ungar and Noy-Meier attempts, in a basically mechanistic fashion based on bite dynamics and from first principles, to predict instantaneous intake rates. Forage quality effects are not included. The GRAZE model is considerably more detailed and complex and functions as a stand-alone simulator of the plant-herbivore community grazing system. However, it is derived from earlier models which may have been constructed with objectives that are not always entirely compatible with those of GRAZE itself. GRAZE is also rather complex, requiring values to be specified for around 110 input variables. FORAGE also seeks to link a pasture productivity model with an animal productivity model but functions in a more transparent, stand-alone manner.

Concentrating, to a greater extent on mechanistic relationships concerned with digestion in the animal, Illius and Gordon have presented a model that integrates the processes of intake and digestion, essentially predicting the voluntary intake of a range of forages from the animal's body size and the degradation characteristics of the forages consumed. Whilst this model did not include a treatment of diet selection, it was able to explain between 61 and 70% of the variation in forage intake over a range of body weights. This would suggest that the approach used might be assimilated into more general pasture and range productivity models in future.

The available literature suggest that there is still some way to go in integrating the ecological and physiological factors that, in plant communities, affect forage production and quality and, in animals, the utilisation of ingested feeds. If models are to provide robust quantitative predictions of the productivity of pastures and range, under a wide range of circumstances, considerable progress will have to be made in this area.

Degradation and digestion of ingested feeds

The delivery of nutrients to ruminant tissues via the processes of degradation and digestion have been simulated in some detail using dynamic modelling techniques.

The relatively simple model of France *et al* established an approach to modelling the separation of ruminal and post-ruminal digestive processes and the interdependence of nitrogen and energy utilisation in the rumen. It also defined a standard for clarity in reporting the construction of simulation models. The basic approach used by France *et al* was developed to a much greater degree of detail by Baldwin *et al* in a model which used 12 state variables to predict the delivery of ammonia, volatile and saturated fatty acids, glucose and amino acids. Examples of other models that have refined the same general approach include:

— Dijkstra *et al* included improved treatments for microbial recycling and substrate preferences, pH effects on microbial activity and ammonia and volatile fatty acid (VFA) absorption and attempted to assess the principal factors affecting microbial chemical compositions.

— *Danfaer*. This model integrated digestive and metabolic processes in the dairy cow using 77 state variables. Due, perhaps, to its size, this herculean effort has yet to be fully validated.

When operated to address the objectives formulated by their developers, a "reasonable measure of agreement" is often, and quite justifiably, claimed for these models. Prediction errors averaging between 15 and 50 % are common. However, it is generally accepted that a number of outstanding problems remain in the prediction of nutrient supply even when the ruminant digestive system is modelled at this mechanistic level:

— Reliable treatments for predicting the stochiometry of VFA production have proved elusive. Most models have resorted to an empirical approach.

— The rumen is not a steady state system (particularly on small farms in developing countries). Alternative mathematical techniques are needed to simulate pulsatile substrate inputs. Adapting existing models for use with pulsed feed inputs has produced disappointing results

— Solid and liquid phase passage rates have not been effectively simulated by mechanistic modelling. Again an empirical approach is usually relied upon.

The taking of a more mechanistic view of nutrient and energy transactions does have some attractions for use in developing countries. Static parameters that are applied routinely in estimating the nutritive value of temperate feeds often fail when applied to tropical feeds. For example, serious under-predictions of animal performance have been widely observed when the metabolisable energy system has been applied to tropical feeds.

Dijkstra *et al* constructed a model of the degradation and digestion of diets based on sugar cane which offers good agreement with experimental observations. However, it is unlikely that this approach will be applied generally, at least in the near future, as its requirements for data that describe, amongst other things, the dynamics of nitrogen, neutral detergentfibre and starch breakdown in the rumen are unlikely to be met for more than a few, intensively-studied feed resources.

Dynamics of Livestock Holdings

The dynamics of livestock herds and holdings— essentially describe changes in their sizes and structures with time. These changes may be introduced directly by the manager of the holding as, for example, when animals are bought and sold. They may occur on an *ad hoc* basis on farms in developing countries but, in some cases, a more systematic culling and replacement policy may be operated. In the livestock systems modelling context, such management decisions would generally be represented amongst the input variables for the model. Indeed, an examination of the consequences of different, defined culling strategies - on holding dynamics in the longer term or on resource generation - may be amongst the objectives of applying this type of model.

Not all changes in the dynamics of livestock holdings arise directly from deliberate changes in management. Figure 3 suggests a schematic representation of the ways in which animal factors may interact to induce variation in the sizes and structures of livestock holdings. Changes in reproductive rates and the incidence of disease may be regarded as the driving factors in this simple model. However, these interact, not only with each other, but also with other factors, most notably the animal's nutritional status that is ultimately mediated by the balance between the size of the holding and the feed resources that are available.

Furthermore, management factors such as the ability to control disease or efficiency of reproductive management may also influence these influences indirectly. This separation of deliberate management strategies and the effects of more subtle and progressive interactions implies a dual role for modelling in relation to herd dynamics:

— an appreciation of the factors affecting livestock holding dynamics may be used to study the impacts of holding management on other aspects of the system such as resource generation and processing;

— an examination of the nature of the impacts of other factors (such as nutrition or disease status) on aspects of herd dynamics may form an objective in itself.

Modelling reproductive performance

A number of models of reproductive performance in livestock holdings have been constructed with the objective of predicting various aspects of reproductive performance. Azzam *et al* describe a stochastic, dynamic model of reproductive performance in beef herds based on the interplay between a number of reproductive events. This model does not implicitly account for variation induced by changes in nutritional status or disease incidence although their implications could be assessed using the model if their effects on its reproductive parameters were known. A deterministic model reported by Werth *et al* uses the outputs of the model of Azzam *et al* to evaluate the consequences of changing reproductive management strategies on net returns from the herd.

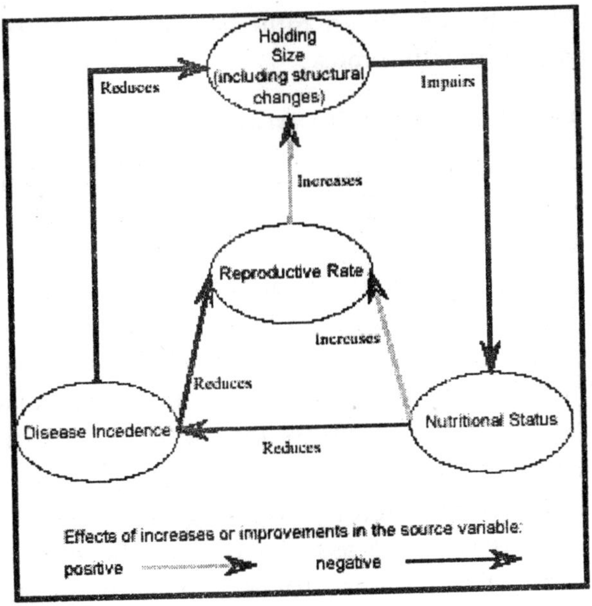

Figure 3. The principal interactions associated with the dynamics of livestock holdings

The TAMU beef model and a number of variants that focus in more detail on aspects of reproductive performance adopt a more deterministic approach. Probabilities of oestrus and conception occuring are calculated from a series of equations adjusted for factors such as body condition, current weight changes and state of maturity. A similar approach is taken in the model of Denham *et al.* Kahn and Spedding have adapted the TAMU equations to allow a stochastic treatment of reproductive events and mortality.

Modelling disease incidence

A number of examples of the mechanistic modelling of livestock parasites and diseases, do exist. Where livestock systems models have been used to study disease impacts, this has usually been achieved by specifying the effects of infection on herd structure directly or on other parameters that exert an influence over herd structures such as reproductive rates.

Resource Generation and Processing

With a reasonably robust treatment of resource utilisation at the individual animal or herd level and the dynamic factors affecting the structure of the herd, a livestock simulation model should be able to generate information on the impacts of changes in these on the generation and processing of resources by the herd. This information may then be used in conjunction with an appreciation of farmers' objectives to assess the suitability of different management options.

The principal factors affecting the generation of livestock products are associated with the efficiency of conversion of ingested nutrients; i.e. with the animal's metabolism. This may be affected by genotype, environment and by ultimate use. In addition, various feedback factors associated with the degradation and digestion of the feed may affect the metabolic process that govern the assimilation of ingested nutrients into productive outputs.

Prediction of long-term body weight changes

For practical rationing purposes in industrialised countries, simple factorial approaches are widely used to predict the demand for nutrients and energy for the maintenance and growth of immature animals. It is recognised that the efficiency with which nutrients are used for different purposes will vary. For example, AFRC have computed effective efficiencies of utilisation of metabolisable protein of one and 0.59 for maintenance and tissue growth respectively. These treatments have been adopted and adapted for use by a number of production systems oriented models. The approach has the attractions of simplicity and availability of input data describing a variety of feeds for running the models.

However, applying the algebraic transformations required to adapt these rationing systems for the prediction of performance is questionable where their constituent equations have been derived using least squares or non-linear regression techniques. Nonetheless, as more detailed, mechanistic attempts to predict growth responses must, in some way, account for their sigmoidicity, these tend to be implemented at a cellular level which restricts their utility in the simulation of agricultural systems.

Production of lactating animals

Similar, simple factorial models are available for dairy animals but their use for predictive purposes is complicated by their inability to predict the relative partitioning of nutrients between different functions, principally to milk synthesis and to short-term liveweight gains. Furthermore, and particularly during early lactation, there may be a release of stored nutrients through the catabolism of body reserves which augments the supply of digested nutrients. Most simulation models implemented at the holding level attempt to avoid this issue by specifying partitioning factors for the particular circumstances under whichthe model is to be run. Again, more detailed models which attempt to predict this partitioning are implemented at the cellular level which makes them unsuitable for incorporation in management models.

Role of livestock in nutrient cycling

The production systems of developed countries tend to focus on the assimilation of ingested nutrients for productive purposes unless those that are voided by the animal in faeces and urine present a waste disposal problem. In the mixed farming systems of developing countries these materials are often regarded as a valuable resource for crop production so models of these systems will be needed that can assess the impact of livestock via these pathways.

The treatment of animal excreta by many soil - crop models is minimal and generally static in nature. Characterisations of manure quality are crude, usually based on gross ratios of C : N which, in ignoring the form in which N is present, may distort sensitivity analyses conducted to examine the impacts of manure use in the systems under study. The current efforts of Probert and his colleagues to introduce improved characterisations of manure into the APSIM modelling environment are noteworthy although these still avoid basing this characterisation on dynamic processes in the animal by specifying manure quality parameters as input variables.

Dewhurst and Thomas suggested an approach to modelling N flows from livestock herds in order to examine waste management implications. This approach was adapted and expanded by Thorne to examine the impact of livestock management on the cropping component of a production system using a similar animal model linked loosely to the SCUAF (Soil Changes Under AgroForestry) model of Young and Muraya. This approach proved promising and led to the development of ANORAC. ANORAC is a fully integrated, dynamic model based on the earlier model of Thorne and the CENTURY soil submodel. It predicts mineralisation patterns in response to changes in organic matter management and differences in animal feeding strategies.

Modelling of this aspect of crop - livestock integration in a dynamic fashion is still relatively undeveloped. However, it offers considerable potential for intervention testing in highly integrated crop - livestock systems where the role played by livestock in maintaining soil structure and fertility may be their most important function.

Draught Animal Management

The role played by Draught animals in crop - livestock integration is unique as their primary role is not as a channel for resource flows but more a management tool for farmers.

Nutrient requirements of and nutrient utilisation in draught animals for various purposes have been studied intensively over the last twenty years, notably at the Centre for Tropical Veterinary Medicine in Edinburgh. This information has been widely published and now summarised and is sufficient as a basisfor modelling of draught animal utilisation. Thus far the only attempt to synthesise the information within a modelling framework appears to be that of Van der Lee *et al* in attempting to develop an animal traction module for PC-Herd. The validation of this model suggested, once again, a reasonable degree of correspondence with observations. However, a number of problems in applying draught animal models may limit their widespread use in future:

— the productivity of draught animals is largely limited by feed energy availability. There are practical problems in assessing this reliably and routinely for tropical feeds;

— impacts of disease on productivity may be of particular significance where the supply of draught is concerned. Some information on these impacts exists in the literature but this may not be detailed enough or adequately describe interactions with nutritional status to allow a reliable treatment to be formulated

— Much of the experimentation has been conducted at a detailed level. Data availability may be inadequate for parameterisation across a wide variety of field problems;

Optimisation and Decision Making

The models describe thus far have been used to describe - at whatever level of detail - the livestock component of agricultural systems and its outputs. In the context of applying management models it is likely that an element of decision making required amongst the predictions of the model for different scenarios will be required. This may be carried out on a more-or-less *ad hoc* basis provided that the user has some appreciation of what farmers' objectives might be in applying the scenarios tested.

Alternatively, the representation of these objectives in terms of constraints and available resources may be formalised so that a more systematic approach to the identification of optimal strategies may be taken. Most optimising models of agricultural systems apply mathematical programming algorithms in order to identify combinations of inputs that may be most suited to a particular set of circumstances.

Mathematical programming

The mathematical programming methods that have been applied to agricultural problems include:

— *Linear programming* (LP), which will identify the combination of input variables that optimises the value of a single response variable. There is a wide variety of applications of LP in farm planning. Multiperiod linear programming allows a dynamic element to be introduced into the use of LP by running repeated iterations with different values assigned to input variables.

— *Multiple criteria decision making* (MCDM) may be based on a range of mathematical algorithms including simplex or others. In developing countries, farmers rarely seek to fulfil a simple objective so MCDMmethods offer a considerable advantage over LP in the analysis of their systems.

— *Dynamic programming* has been less widely applied due, perhaps largely, to difficulties in constructing the relatively complex model specifications that it requires.

Artificial intelligence methodologies

More recently, a number of methods developed by researchers in artificial intelligence (AI) have become available to the systems analyst. Some of these have considerable potential for application in the optimisation of management strategies for agricultural systems. Their advantage is that, in being based on less rigid criteria specifications, they may generate a wider variety of potential solutions, some of which may have been excluded by the application of mathematical programming methods. This, of course, means that the need for informed interpretation of their outputs is even more crucial than with the application of more traditional optimisation methods.

Examples of the application of AI methods to the identification of solutions to agricultural problems are not easy to find currently. The two most promising techniques would appear to be:

— *Genetic algorithms*. These can, in theory and if correctly formulated, search a much wider space for optimal or near-optimal solutions than mathematical programming methods.

— *Neural networks* are widely used in pattern recognition applications. They are suited to problems which exhibit a high degree of parallelism so may be of use in for solutions in highly dynamic situations where objectives change regularly.

Integrated Crop-Livestock Models

A number of livestock models are, potentially, available as a basis for the development of integrated models of crop - livestock systems. These operate at a range of different levels of detail. However, they generally focus on a single species or type of production system and are not, individually at least, that well suited to simulating the multi-species holdings that are common in developing countries. It is, therefore. unlikely that any single model could meet all of ILRI's requirements for modelling crop - livestock systems.

Many of these models are, in fact, quite closely related to each other with some, initially stand-alone models, having been adapted and assimilated into broader or more detailed models. The use of TAMU beef derivatives by SPUR offers and example of this.

TAMU Beef

Language:	FORTRAN 77.
Platform:	Any with an appropriate FORTRAN compiler. Source code freely available to interested researchers.
Cost:	Free.
Strengths:	Long pedigree during which it has been widely used with proven results; Even though the model is quite highly aggregated in terms of timestep and the grouping of animals into classes it is sufficiently generalised to simulate a wide range of optional scenarios including the effects of disease;
	The input data set is simple but, again, this does not seem to hinder the model's use under a wide range of circumstances.
Weaknesses:	Long pedigree during which many of its treatments of animal processes have been refined elsewhere. To an extent, some of the derivatives have compensated for this;
	Feed intakes are effectively specified amongst the input variables although, again, some derivative uses improve on this considerably;
	The characterisation of nutritional inputs in terms of dry matter and crude protein only would not allow the effective evaluation of nutrient cycling issues.

Overview

Texas A & M University's general cattle production systems model is probably the earliest, deterministic livestock systems simulation model to have found widespread use for the evaluation of practical production problems and the identification of potential solutions to them. It has been applied, not only in its original form, but also with later modification and in association with, for example, more detailed models of elements of the range / pasture - livestock ecosystem.

The primary objective of the developers of TAMU beef was to provide a model that would be able to predict performance levels of different animal classes within the herd from both their genetic potential and feed inputs. The model is able to simulate performance under a wide range of conditions relating to breeding seasons, herd management policies and feeding regimes that result from changes in management and from environmental variation. A large number of modified versions of TAMU Beef exist including the model, described in brief, by Cartwright *et al* which allows the milking of cattle practised in dual purpose systems to be simulated.

Level of detail

The original TAMU beef model operates on a monthly timestep. The model does not allow the simulation of individual animals. Instead, animals of different ages and at different stages of the production cycle (e.g. calves, heifers, pregnant cows, lactating cows) are aggregated into a range of groups which are re-populated dynamically as a simulation run proceeds. Reproductive events and mortality are treated deterministically in relation to the time of year (season) current nutrition and body condition.

A number of the derived models offer improvements in respect of level of detail although these may come at the price of useability in terms of the availability of input data. Typically, the model of Kahn and Spedding allows a timestep of between 1 and 30 days to be specified by the user, groups of animals may be disaggregated to the individual animal level and the randomly occurring, discrete events associated with herd dynamics are, more properly, treated stochastically.

Validation and calibration

TAMU beef was widely tested for a range of conditions and production systems, including multipurpose cattle, (in Botswana, Colombia, Guyana, Tanzania, the USA and Venezuela) during its development with the authors claiming close coincidence with observed levels of production. It has been suggested that the model is best suited to situations in which only poorer quality pastures are available as predictions of digestibility are based on N content. Seman *et al* observed systematic over-prediction of liveweight with the modified TAMU beef model that forms part of the SPUR model under a scenario in which forage availability and N supply were not limiting. In many situations in developing countries this may not necessarily represent a problem.

Examples of existing applications

Sullivan et al linked the original TAMU beef model with a forage production model to study grazing systems in East Africa. The forage model allowed the generation of profiles of seasonal feed availability from climatic variables (rainfall, evaporation and irradiation). However, estimations of forage quality were based on tabular values for average monthly crude protein contents and digestibility values derived from a secondary source.

Validation of the model was conducted for a village system in Tanzania with accurate estimates of reproductive and mortality rates and growth performance being achieved.

The model was used toexamine the impacts of changes in management systems from "traditional" management include shortening the breeding season, early weaning, supplementary feeding and managing herd sizes to account for seasonal changes in feed availability. It was able to derive a number of management recommendations based on the outcomes of these simulations. The forage production component of this model would probably be regarded as crude now but the work is notable as an early attempt to model a system across the disciplinary boundaries of its component parts. The model has also found widespread use in support to the extension services both in Texas and other areas of the United States. These include the evaluation of breeding policies, impacts and cost benefits of disease control interventions via their effects on herd dynamics and the effects of subsidies on the profitability of the beef enterprise.

As stated earlier the general structure and approaches used in TAMU beef have found widespread use in other modelling exercises. The treatments applied to stage of maturity and its interactions with nutritional status, both current and past, have had at least some influence, directly or indirectly, on most of the other herd scale models described here.

Potential for future applications

In spite of its age, TAMU beef should not be discounted entirely. It is available in a robust FORTRAN implementation that is transportable amongst platforms. It should prove possible to derive its relatively simple input dataset for a wide variety of circumstances. It suffers from a lack of detail that means that it would be difficult to apply in studying the more subtle aspects of crop-livestock interaction without major modification. However, for a quick and dirty approach to evaluating the impacts of, for example forage management and supplementation strategies on herd dynamics, TAMU Beef might offer the most economical solution.

JAVA / PC-Herd

Language:	Turbo Pascal
Platform:	IBM PC and Compatibles.
Strengths:	Interactive construction allows step-by-step analysis of management strategies;
	PC-herd includes the only detailed treatment of draught animal management and use currently available.
Weaknesses:	Interactive construction makes conduct of lengthy simulations laborious;
	Level of detail in feed resources dataset may make the model's use impractical in many circumstances.

Overview

JAVA and PC-Herd were both developed in the Department of Tropical Animal Production of Wageningen Agricultural University during the late 1980s. Both models focus, primarily, on the mediation of animal performance by fluctuations in the availability and utilisation of feed resources. They were developed with the aim of providing tools for the analysis of year-round feeding strategies as part of the department's project work, in Indonesia and Bangladesh.

JAVA may, in essence, be regarded as a prototype of PC-Herd which is able to simulate herd dynamics and the range of outputs of a livestock holding at a more mechanistic level. Indeed, PC-Herd appears to be unique at present in integrating a detailed, mechanistic draught animal feeding and management module.

Level of detail

JAVA is a relatively simple model which does not attempt to simulate performance mechanistically. Instead requirements for maintenance and gain are specified amongst the input data and performance is then assessed against the energy supplied by the available feeds allocated according to an intake function selected by the user.

PC-Herd is of modular construction with feed supply, herd dynamics and animal traction modules that may be used individually or together. The programme may be run in two modes, simulating a grouped herd or at a more detailed level allowing the behaviour of a herd of individually characterised animals to be simulated. The latter option requires data that will allow the prediction of individual feed intakes but allows more detailed outputs of feed intake, energy requirements, energy balance, weight changes, optimal weights, milk yields, draught use, manure production, general reproductive parameters and mortality may be produced. Herd dynamics are treated stochastically. Timesteps for both models may be varied from one day upwards.

Input data

A detailed summary of PC-Herd's required dataset is not available at present. It is similar to that of JAVA in respect of its description of feed resources but is augmented with information at a greater level of detail on animal characteristics and general herd management. Both models allow a very detailed characterisation of patterns of availability of feed resources and are relatively strong on the simulation of stall feeding situations. However, this carries with it the overhead that, for the timesteps of less than one month that are likely to be necessary for the model to be sensitive to data of this level of detail, the effort that must be put into collating a feed calendar to run the model may be excessive.

Validation and calibration

Both models have been fully but not widely validated. JAVA has been tested against field data collected in Indonesia and PC-Herd for a mixed farming system in Bangladesh.

Examples of existing applications

JAVA and PC-Herd have not had widespread use with datasets other than those for which they were originally validated. The models are not currently being promoted by their developers although they are freely available to interested users. One likely reason for this relative lack of use would appear to lie in the difficulty of generating comprehensive data sets to describe feed availability.

Potential for future applications

Whilst PC-Herd has not received widespread use, it may have considerable potential where detailed, dynamic data describing feed utilisation are available. The joint ILRI / NRI project on feeding strategies for dairy cattle is generating such a dataset for a group of study farms in Kiambu District, Kenya.

Overview

The modelling approach described by Sørensen *et al* essentially links instances of a dynamic mechanistic model simulating the individual dairy cow or heifer

(SIMCOW) with a stochastic dynamic model (SIMHERD) to provide an aggregated view of herd dynamics.

SIMCOW / SIMHERD

Language:	Turbo Pascal
Platform:	IBM PC and Compatibles
Strengths	The individual animal and herd models are well-integrated and appear quite sensitive to the effects of interactions between processes at the individual cow and herd levels;
	A good compromise has been struck between the flexibility to simulate a wide range of managerial interventions whilst not requiring an impossibly complex input data set.
Weaknesses	The lack of a treatment for N transactions in the single cow model may limit the model's utility in simulating crop-livestock systems.
	How meaningful the stochastic, dynamic treatment of herd dynamics used in SIMHERD would be for the very small herd typically found in crop - livestock systems is unclear;
	Treatment of energy transactions based on Scandinavian Feed Units. Limited data would be available on tropical feeds although it should be possible to estimate SFUs from the parameters used by other systems.

The SIMHERD model has been used in other modelling activities including the Edinburgh model and studies reported by Plaisier et al. The latter case, included a more detailed treatment of reproduction and milk production in the SIMHERD model and the replacement of the SIMCOW individual animal model.

Level of detail

SIMCOW provides a net energy based simulation of production responses to feed utilisation at the individual cow level. It assumes that requirements for maintenance and pregnancy are always met, either from current intake or from the mobilisation of body reserves. Energy supplies above this are partitioned between milk and body reserves according to potentials for milk production and growth and a partitioning rule. Milk production potential is derived from the gamma function of Wood. As the general applicability of this approach in smallholder dairy herds has been questioned by Tanner et al it would probably be appropriate to derive alternatives if SIMCOW is to be used in ILRI's modelling activities. SIMHERD simulates discrete events, including heat detection, conception, foetal death, sex and calf viability, stochastically. Herd size is restricted to a user specified maximum and a number of rules relating to culling policy are hard wired into the model in order to ensure that this is not exceeded. This aspect might require modification if the model were to be used in simulating smallholder mixed farming systems.

Validation and calibration

The feasibility of a full validation of the integrated models is discounted by the authors on the grounds that it would be "impossible to obtain data over the number of years necessary for thorough statistical analysis". For this reason, potential users are encouraged to validate sets of production parameters for individual animals within the SIMCOW model before attempting to use it with SIMHERD for simulating herd dynamics. This strategy has been used by the developers of SIMHERD / SIMHERD who quote good levels of agreement with published data in responses to changing feeding strategies for different types of cow. An alternative approach to validation is described by Enevoldsen and Sørensen who state high levels of user acceptance of SIMHERD used for decision support in planning health management strategies.

Examples of existing applications

SIMCOW and SIMHERD have been quite widely applied both by the developers and their colleagues and as components of other simulation models. A number of these applications have been described, in detail, in the literature:

- Kristensen et al studied the effects of grazing intensity on individual production using SIMCOW and on herd production using SIMHERD for the management conditions of twelve commercial dairy herds in Denmark. The modelsproved sensitive enough to differentiate the general adverse effects of high grazing pressure on milk production and body weight changes from herd specific effects on nutrient partitioning to milk or body reserves.
- Ostergaard et al used the models to evaluate the interactions between the form of feeding of a total mixed ration and other herd management interventions. Effects on production and economic returns were assessed and found to be explained largely by the effects of diet on the intake of the individual cows.
- Sorensen et al examined the implications of changing dry period lengths on production levels and economic returns. The model was able to identify and economic optimum for the dry period under a range of culling strategies and feeding strategies although outcomes were not particularly sensitive to the management changes tested.
- Enevoldsen and Sorensen and Sorensen and Enevoldsen describe the use of SIMHERD in planning dairy herd health management strategies.

— A number of applications in simulating the physical outcomes and economic benefits of different reproductive management strategies have been reported by Plaisier et al.

Potential for future applications

SIMHERD is now a widely used model that is implemented with a relatively userfriendly interface. It does have some limitations for predicting impacts in other system components but these arise mainly from the nature of SIMCOW (the individual animal model). These have been successfully circumvented by the replacement of SIMCOW in some of the derived implementations. The model seems to be establiching itself as something of a standard approach to modelling herd dynamics and, thus has the greatest potential for application to problems implicated, directly or indirectly, in influencing herd sizes and structures.

SPUR - Simulation of Production and Utilization of Rangelands

Language:	FORTRAN
Platform and operating system:	IBM PC and compatibles / MS-DOS
Strengths	A powerful integrated model of plant and animal production in plant-herbivore systems; Where input data are available, SPUR offers a powerful, process-based tool for simulating plant-animal communities
Weaknesses	Detailed input data set may not allow parameterisation for a wide range of situations in developing countries.

Overview

SPUR is a general, grassland ecosystem model that is flexible enough to allow the simulation of wide range of plant-herbivore communities including heterogeneous grazing environments and both wild and domestic herbivores. The original objectives in developing the model were to:

- — evaluate rangeland systems and provide a basis for management decisions;
- — optimise rangeland management systems for desired multi-use products;
- — plan and evaluate land improvement practices;
- — provide a computational basis to investigate the impacts of environmental modifications on alternative management strategies, and;
- — forecast the effects of climate change on range ecosystems.

Clearly, the model could have considerable potential for application in systems in which domestic animals have significant access to natural grassland communities! The original SPUR I model has four components; hydrology; plant growth; animal physiology and production and economics. The operation of the model is based primarily on responses to changes in driving environmental variables including precipitation, maximum and minimum temperatures, solar radiation and wind speed.

The revised SPUR II model predicts beef cattle performance and production by simultaneously simulating the production of up to 15 plant species on 36 heterogeneous grassland sites. These simulations are based on the hydrology, nitrogen cycling, and soil organic matter turnover of grazed ecosystems as well as production from rangeland under different climatic and environmental conditions, and management strategies. The latest version of SPUR that does not appear to be widely available yet includes a improved cow-calf androoting depth models as well as the CENTURY model of Parton et al n order to enhance the accuracy of prediction of organic matter turnover in the system.

Level of detail

SPUR II simulates carbon and nitrogen cycling through aboveground live and dead vegetation, live and dead roots, litter, and soil organic matter. Effects of stocking rate, grazing season and system are also incorporated. The model simulates the direct effects of CO—on plant production and therefore may be applied in climate change studies. Driving environmental variables may be derived directly from existing meteorological records or using an integrated, stochastic weather generator. The hydrology component calculates upland surface run-off volumes, peak flow, snow melt, upland sediment yield, and channel stream flow and sediment yield.

Soil-water tensions, used to control various aspects of plant growth, are generated using a soilwater balance equation. Surface run-off is estimated by the SCS curve number procedure and soil loss is computed by the modified universal soil loss equation. The mechanistic, process-based structure of SPUR II makes the model well suited for examining the interactions between management decisions and climatic influences on short-term ecological processes and evaluating possible adaptive management strategies. The model may be run to simulate changes over periods of up to 100 years.

The animal component of SPUR II is based on the "Colorado Beef Cattle Production Model" (CBPCM) that is a modification of the original TAMU Beef Model. CBCPM is a herd wide, life cycle simulation model and operating at the individual animal level that allows a variable time step of 1 to 30 days; a variable herd size;

deliberate managerial manipulation of herd structures, the ability to simulate the effects of different crossbreeding systems and the evaluation of the effects of genetic selection in the longer term. CBCPM's biological routines simulate animal growth, fertility, pregnancy, calving, death, and demand for nutrients. The impact of genetic traits related to growth, milk production, fertility, body composition, and survival can also be studied.

Intake at grazing is calculated by FORAGE, a deterministic model that operates as an interface between CBCPM and the SPUR II plant-soil model. FORAGE is driven by bodyweight derived from growth curves, demand for forage, and the quantity and quality of available forage at each time step in a simulation. Intake of grazed forage is determined by simulating the rate of intake and grazing time of each animal in the time step. Forage removal by a variety of wildlife species may also be simulated using SPUR II.

Input data

SPUR II is a detailed, process-based model integrating plant and animal communities at a consistent level throughout. Consequently, its requirements for input data are detailed and may be difficult to meet for the circumstances found in developing countries. More than 1100 state and decision variables are listed for the animal component of SPUR II alone.Validation and Calibration The main thrust of validation of SPUR has been conducted with data describing rangeland production collected over a 30 year period on three ranches in the Rolling Plains of Texas. The formal validation of the animal component has used animal production data from a series of experiments conducted at the Texas Agricultural Experiment Station. In its current state of development, the authors suggest that the most pressing improvements required to the model lie in its hydrológical routines and in the addition of routines for simulating the production and utilisation of brush species. Both of these issues are likely to be of importance if the model is to be used in simulating production systems in developing countries.

Examples of existing applications

Perhaps due to its complexity, the SPUR models have undergone a long development phase and only just appear to be reaching some kind of maturity. Thus reports of their use are relatively sparse. Baker et al have used the model to evaluate the likely impacts of various climate change scenarios for production levels from the US range beef industry.

Potential for future applications

SPUR is the most disaggregated model considered here and this is likely to have considerable implications for its potential utility. Where data are available to parameterise the model, it is likely to be a powerful tool. However, even the SPUR's developers accept that:

- — The model requires fairly detailed parameterization when new plant species are added to the system;
- — CBCPM (Colorado Beef Cattle Production Model) is a very detailed genetic model and may be to complex for many of the range management questions under consideration.

It is suggested that SPUR would be the model of first choice where attempts are to be made to assess crop-livestock interaction across a rangeland boundary. However, where such activities are being developed, data requirements and availability should be given particularly careful consideration at the planning stage.

7

Crop-Livestock Integration in Slopelands

Farm families are finding it increasingly difficult to earn a livelihood from slopeland agriculture because of stagnating yields, the increased cost of inputs, adverse and disorganised markets, and growing population pressure. However, scientists are working to help improve slopeland agriculture in the region.

Farmers tend to integrate livestock and crop production better than most scientists do. Traditional systems already combine various mixtures of livestock species and crops. The challenge is to increase the productivity of traditional systems, so they produce a higher usable biomass while conserving the natural resources on which the whole agricultural system depends.

Crop-livestock integration increases the overall productivity of ecosystems, since livestock can use otherwise unusable parts of plants and crop by-products. Conversely, crops can utilise the livestock manure as a source of plant nutrients. Crop-livestock integration also increases sustainability. It is the flow of energy and materials that determine the stability of slopeland ecosystems. Closed systems in which nutrients and energy are recycled are much more sustainable than an open system which leaks nutrients, and which needs higher inputs to maintain its productivity in the long term.

The most productive ecosystems are multi-layered and diverse. It is this type of ecosystem which produces the highest biomass, and does most to conserve soil and water resources. Most of the models incorporate perennial crops, either for cash or for forage. Leguminous trees and shrubs were particularly important. They can be used as a source of livestock feed, and can also fix atmospheric nitrogen and thus increase soil fertility.

In traditional slopeland farming systems, cattle and goats are often taken up onto higher ground away from cropland during the wet season, or left to graze in the forest. After harvesting, the animals are brought back to the village to graze on the fallow croplands over the dry season. Although farmers in principle are using separate zones for crops and livestock over the growing season, in practice there is often competition between the two. Grazing animals sometimes damage crops, and farmers may have to put a lot of time and effort into protecting their crops with fences.

Cut-and-carry livestock production systems are well suited to Asia's intensive farming. In the Sloping Agricultural Land Technology (SALT) system, developed in the Philippines, goats or other small livestock are fed on nitrogen-fixing legumes grown in a forage garden, usually together with food and cash crops. Leaves and branches of high-protein species such as Desmodium rensonii, Gliricidia septum, Indigofera tycsmani and Leucaena leucocephala are cut periodically and used as livestock feed for penned animals. This system gives more efficient land use and higher milk production than free-ranging animals. The fact that the animals are penned also makes handling easier, and allows the farmer to closely monitor the health and breeding cycles of his livestock. Pens also ensure that livestock manure is deposited in a limited area, so it can be collected to use as fertiliser.

Disadvantages include the heavy labor demand, the possibility of infectious diseases among penned animals, and possible adverse effects when high levels of some legume shrubs are used as feed. For example, some leucaena leaves have a high mimosine content, which can cause hair loss and poor performance in swine. Other species of leguminous shrub can sometimes cause bloat in sheep and cattle. While cut-and-carry livestock systems are more intensive and potentially much more productive than free grazing, farmers need to apply a higher level of management and different skills.

The traditional system of crop production on slopelands in tropical Asia is slash and burn, in which forest is cut and burned for the production of annual food crops. The cleared land is abandoned after a few years, and the group moves on to clear new land. In the past, when populations and land clearings were both small, the forest could be allowed to regenerate for thirty years or more, giving a sustainable cycle. As population pressure has risen, however, cultivators are clearing larger areas of land, and leaving short fallows of only a few years. As a result, yields are falling, putting even more pressure on available land.

Weeds are a major constraint. Slash and burn farmers nearly always rely on manual rather than chemical weed control, and shorter fallows mean more weeds. One solution is the accelerated fallow, using legumes and integrated livestock production. Another solution is the extended fallow, based on perennial cash crops. This also often integrates livestock into the production system, using trees such as paper mulberry as forage. Rotations with improved fallows may help to overcome the serious weed problems, while livestock will benefit from access to high-quality feed resources.

Development of Slopeland Technology

An important new development is participatory approaches to technology development (PTD). The process of PTD begins with a meeting of researchers and farmers, where farmers identify their main problems, give them a priority, and describe what they have done in the past to solve them. Researchers then bring to farmers useful information, planting materials etc. which they think will benefit farmers.

Farmers test these for themselves in whatever way they prefer. When they develop promising technologies, scientific experiments may be carried out to validate the farmers' experience. Once farmers are testing technologies, there follows a period of evaluation in which farmers describe what technologies they like and why they like them, and what features need to be improved. Evaluation identifies, not only promising technologies, but also the criteria being used by farmers to judge whether a technology is successful or not.

The integration of crops and livestock is a promising agricultural system for low-income small-scale farmers. A great advantage of crop-livestock integration is that it uses diverse resources such as fodder legumes, crop residues and livestock manure in a system of nutrient recycling. Livestock provide a high level of profit per unit of labor input, plus valuable manure for use as fertiliser. Livestock also have the great advantage of being relatively easy to market compared to harvested crops. There is a steady demand for livestock products, with relatively high and stable prices.

Another great advantage of livestock production in slopeland agriculture is that the animals can walk to market. This is a considerable benefit in remote areas with poor roads, where farmers would otherwise find it very difficult to market what they produce. In such areas, livestock may be the only feasible source of cash income.

The way forward seems to be highly productive mixtures of species, including a range of complementary crops at different levels in the canopy, and forage species as livestock feed. At present, many systems of crop/livestock integration are based on the use of separate zones during the cropping system, and free grazing in fallow fields during the off season.

Such systems are already giving rise to conflict, and damage of crops by livestock. Such conflict is likely to intensify as populations become even denser and the cropping period is extended. They will probably be replaced by cut-and-carry livestock systems based on crops of nitrogen-fixing legumes and high-yielding improved grasses.

Similarly, slash and burn agriculture is undergoing a transition to sedentary cultivation. For unimproved systems, this is a process of impoverishment, with progressively lower yields and worsening weed and erosion problems. However, the Workshop presented a number of hopeful examples of how this downward spiral can be halted and reversed with appropriate cultivation methods based on complementary crop and livestock species.

Livestock can also be integrated into slopeland reforestation programs, particularly in the early stages of tree growth. New systems of research and extension based on PTD are helping ensure that technology is more appropriate to farmers' needs, and that it solves those problems which are seen as crucial by the slopeland farmers themselves.

Crop-Livestock Integration Patterns

Many patterns of smallholder croplivestock integration patterns can be seen in various countries.

VAC in the Midlands of Vietnam

The components of this crop-livestock integration are a garden ("Voun"), fishpond ("Ao") and livestock pen ("Chuong"), hence the term VAC. The term is somewhat misleading, since not all farms have fishponds (Cuc et al. 1990). The VAC is usually integrated with a rice field whenever natural conditions favor a rice paddy system. The garden is diverse, containing fruit trees, hardwoods (Melia sp. etc.), bamboo, vegetables, sugarcane and others. Livestock include cattle, buffalo, pigs and chickens (Fig. 1). VAC is generally described as an integrated and relatively closed nutrient cycling system.

Paddy Rice - Buffalo - Forest Grazing in Thailand

This system has paddy rice in the lower portion of the landscape, and water buffalo which are used for land preparation and draft power, but which are allowed to graze in the paddy fields after harvest. During the ricegrowing period, buffalo are grazed in hilly and mountainous areas. Whether they spend the day in fallow fields or in the mountains, the water buffalo are kept in a pen under or near the house at night. Accumulated manure is used as fertiliser for rice fields or the home garden. Sloping Agricultural Land Technology (SALT), Mindanao, Philippines SALT is a technology developed by the Mindanao Baptist Rural Life (MBRLC), in the southern Philippines. Farms have hedgerows, usually legumes, planted along the contours for erosion control.

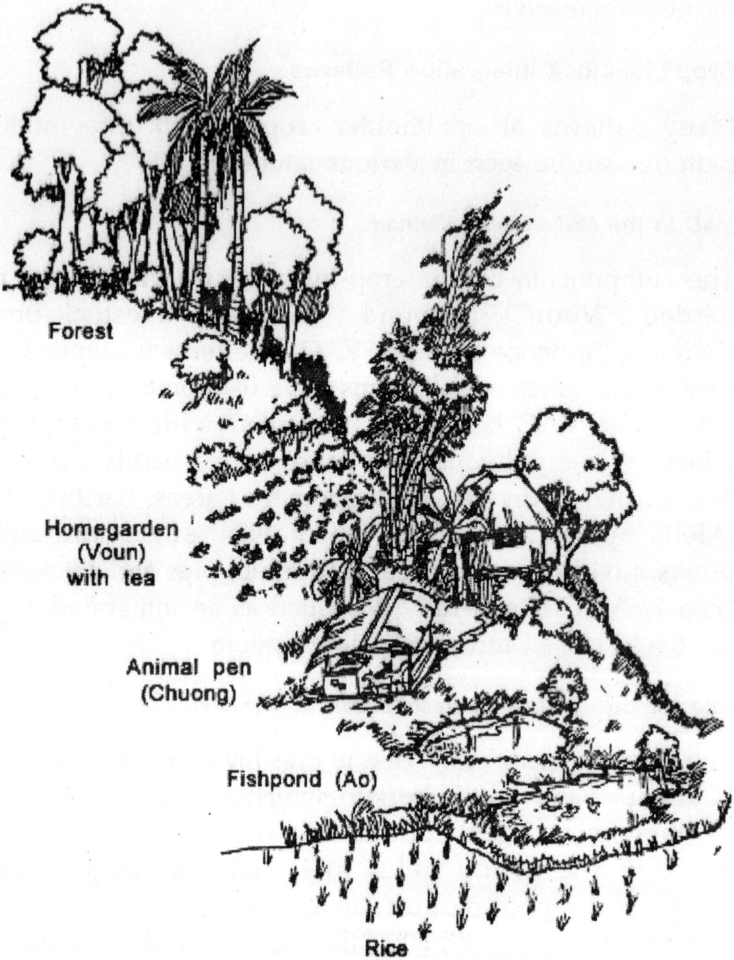

Figure 1. Diagram of a model VAC farm

The hedges also provide forage for goats in a cut-and-carry system. Crops are grown between hedgerows on gentler slopes, and bamboos and trees in steeper areas. The goat manure is used as fertiliser. The goats are raised for meat and milk.

Cattle-Forest/Forage Crop Integration in Central Java, Indonesia

In central Java, at elevations higher than 400 m, cattle are raised in pens near water sources. The cattle are fed on grasses harvested from the forest floor of communally owned pine forest. Their manure is discharged into the water system. They are the major source of income for local smallholders. These are just some of the patterns of smallholder crop-livestock integration found in the slopelands of Southeast Asia. The purpose of this Bulletin is not to describe all patterns of crop-livestock integration, but to use the ones described above to analyse the ecological implications of this type of farming.

Ecological Role of Crop-Livestock Integration

In the context of Southeast Asia and the humid tropics in general, there are two basic ecological principles which must be taken into consideration to understand the implications of crop-livestock integration in slopelands.

1. Devoid of adequate vegetation cover, slopelands in the humid tropics are prone to soil erosion. They suffer from water shortage, and become marginally productive and unstable. The most stable ecosystem in the humid tropics is a tropical rainforest ecosystem, which is multi-layered, diverse, protective of the soil, hydrologically stable, productive and protective of other ecological processes. An ecologically sound pattern of crop-livestock integration in Southeast Asia will be one which mimics this original ecosystem.
2. Crop-livestock integration follows the trophic level structure of ecosystems, where crops represent primary producers and livestock the grazers or herbivores. It therefore follows the regular food

chain, and energy and nutrient transfer processes, characteristic of natural ecosystems (Fig. 2).

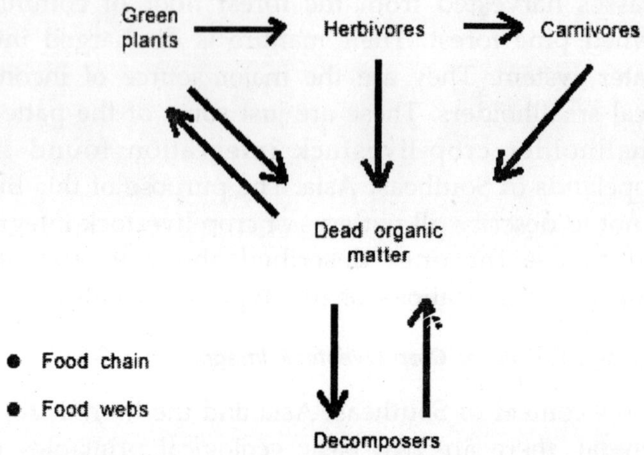

Figure 2. Trophic level niche of livestock

These frameworks help provide a better understanding of the ecological role of croplivestock integration. Some of its most significant ecological functions are as follows.

Determining biomass production

Generally speaking, crop-livestock integration increases the overall productivity of ecosystems. While primary productivity can be attributed to the crop component, livestock as herbivores can add to the overall productivity, especially if they make use of otherwise waste products and indigestible portions of crop plants. Livestock in smallholder farming systems not only increase food production and the protein supply, but also provide raw materials, draft power and manure, and are a risk buffer and a form of savings.

Regulating nutrient and material flow

According to its trophic level and ecological principles of the food chain, crop-livestock integration will determine the pattern of nutrient cycling, as well as the flow of materials and energy in the agroecosystem. The patterns of these flows will determine the stability and sustainability of the slopeland agroecosystem. When these patterns of nutrient, energy and material flow are relatively "closed" and integrated, the agroecosystem is relatively sustainable, compared to one in which these flows are more "open". An open agroecosystem usually causes more pollution, as it is characterised by greater soil erosion and "leaking" of nutrients. It therefore has greater off-site impact, and requires more inputs to maintain its productivity level in the long term.

Determining the level of biodiversity

Crop-livestock integration generally enhances biodiversity in slopeland agroecosystems, compared to monoculture. The level of biodiversity, and the compatibility of the different crops and livestock in the farming system, will influence how stable productivity is on fragile and marginal slopelands.

Serving as biotic carrier for tansfer of diseases and pests

Crop-livestock integration, while regulating the flow of nutrients and materials, also serves as a carrier of biotic elements which may affect the spread of pests and diseases.

Influencing the balance of globally important gases

Crop-livestock integration contributes to the production of globally important gases such as methane. Methane production by livestock is the result of activity by anaerobic bacteria in the animals' gut as they break down organic matter. Ruminant animals worldwide contribute about 15% of the world's methane emission. Cattle and buffalo account for around 80% of the total methane produced by animals.

Ecological Implications if Crop-livestock Integration

Cropping pattern and crop-livestock compatibility

The SALT system developed in the Philippines is an example of a compatible system in which legume hedgerows give soil erosion control and are a source of fertiliser for the crops planted between the hedgerows, and of feed for the goats raised in a cut-andcarry system. If goats were allowed to graze freely, vegetation cover would be removed or reduced significantly, leading to soil erosion and loss of nutrients.

In the case of the Phu Wiang Watershed, Northeast Thailand, the cassava crops planted on the upper slopes of the watershed are a major source of nutrient outflow. This could lead to long-term depletion of soil fertility. In other forms of crop-livestock integration in Southeast Asia, such as a coconut-livestock integration system, cow dung may be a habitat for the beetle that attacks the young shoots of coconut. The crop-crop and crop-livestock compatibility would, therefore, influence the productivity and sustainability of smallholder integrated farming systems.

Nutrient and material flow

Crop-livestock integration in slopelands generally contributes to nutrient and material cycling and conservation. However, data obtained from studies at the Phu Wiang Watershed in Northeast Thailand and the

VAC of upland Vietnam indicate that the pattern of livestock grazing influences the long-term productivity of uplands.

If grazing management consists of livestock being kept in the uplands and transferred to the lowlands for sale, nutrients will accumulate in the lowlands while the uplands are continuously being depleted. This is going to cause a decline in upland productivity in the long run. Dumping of livestock wastes into the river, as is the case of crop-livestock integration in Central Java, causes water pollution. Landscape management should therefore be emphasised.

Spread of pests and diseases

Free-ranging livestock, or even livestock kept in pens, may serve as biotic carriers of pests and diseases. For example, the manure of grazing livestock could serve as a carrier of upland weed species when applied as fertiliser to home gardens or lowland rice fields. Such weeds may become dominant on bunds, and get dispersed into other ecosystems. In the Phu Wiang Watershed in Northeast Thailand, water buffalo infected with liverfluke from Ubon Rat carried this parasite to the Phu Wiang agroecosystem.

Emission of globally-important gases

Methane emissions from livestock, especially ruminants, contribute to greenhouse gases, a probable cause of global warming. The level of methane emitted can be partially reduced by the use of certain chemicals which suppress the activity of anaerobic bacteria in the rumen of livestock, or by integrating nonruminants into the farming system.

Changing patterns of Crop-livestock integration

Crop-livestock integration in Southeast Asia, especially for smallholders, is changing rapidly. The factors responsible for these changes have long-term implications for slopeland productivity. In the uplands of Vietnam, the policy of the government is to promote a market economy, and allocate land to upland farmers. This has enhanced the greening of the uplands, because farmers have been encouraged to plant more fruit trees for the market. This increase in vegetation cover in the hilly portion of the VAC has increased both productivity and soil conservation.

The introduction of electrical power into the Phu Wiang area of Northeast Thailand has allowed families to acquire appliances such as television. These give access to new information about commercial goods, including power tillers (locally known as iron buffaloes). This has resulted in the replacement of many water buffalo by power tillers, which means there is less buffalo manure available as fertiliser.

Since water buffalo are usually tended by young children, this has meant an increase in the free time available to children. They tend to spend this extra free time watching TV or on their studies, depending on the values prevailing in the household. These two cases illustrate how patterns of crop-livestock integration in slopelands can be changed by government policies, market forces and land tenure.

The pattern of crop-livestock integration has strong ecological implications for slopeland areas. It is therefore recommended that the following points should be considered in planning and management.

— The pattern of integration should make sure of crop-crop and crop-livestock ecological and economic compatibility, for optimum biomass production and environmental protection.

— The sustainability of crop-livestock integration should be assessed in view of the nutrient and material flow at the watershed level, and not only at the farm level.

— Crop-livestock integration should promote soil and water conservation, nutrient conservation, and minimise the emission of greenhouse gases, especially methane.

— In developing crop-livestock integration, the possible spread of pests, diseases and polluting chemicals should be kept in mind, and their potential impact on human health and welfare.

— An appropriate Environmental Impact Assessment Framework should be developed for crop-livestock integration in slopelands, of a kind that smallholders can use in their farm planning. This framework should extend to monitoring and evaluation at both a farm and watershed level, to ensure the sustainability of the farming system.

Socio-economic Impacts of Crop-livestock Integration

Most uplands in Asia, particularly the Philippines, face an ecological crisis. The slopeland areas, which account for more than 80% of the uplands, have been subjected to severe natural resource loss due to indiscriminate deforestation and destructive upland farming practices. These activities when combined with natural causes of soil erosion lead to the demise of slopelands or to its unproductiveness. The current rate of exploitation has become so alarming, that many rural development programs have begun focusing their goals on reducing environmental degradation while providing income generating opportunities for the inhabitants and farming communities in these areas.

Despite this critical situation, many agriculturists and environmentalists agree that solutions are available to minimise if not totally eliminate the negative economic, social, and environmental impact of these destructive endeavors. As farming remains the main livelihood source of these inhabitants, there is a need to assist them identify and engage in sustainable farming practices that will help rationalise resource conservation with farm income generation.

One option is crop-livestock farming, a system that continues to expand in the slopeland areas. Just as trees and cash crops are integral to many indigenous farming systems, so too are animals. In conventional agricultural science, animal production is regarded as integrated with crop farming. Animals provide manure and draught power for cultivation, forages are part of the cropping cycle, and the animals and crops form part of the same production unit. Whereas, these forms of integration do exist in smallholdings in the tropics, additional forms of crop-livestock integration have also evolved, occurring not only within but also between production units.

Basic Needs of Crop-Livestock Enterprise

Countries may have different conditions owing to their respective policies and laws governing slopelands. It is recognised, however, that the following basic needs may apply partly or fully for the fulfillment of the potential of this slopeland enterprise.

As the tropics and subtropics can grow and raise almost the same crops and animals, the selection of what crops can be integrated with animals, or what animals can be integrated with crops, will greatly depend on the availability of the following elements:

Land

Slopelands are mostly marginal lands. It needs different types of farming practices to be able to maximise production in a sustainable manner. Considerations in determining the land requirement of a crop-livestock enterprise include:

1. *Suitability of the crop.* The crop to be planted should be the main determinant in selecting the land. For instance, in a corn-cattle project, the adaptability of corn is very important in selecting the area. The residue of the corn will be used as feed for cattle, and when there are no residues, supplementary feeds can be gathered or procured. Corn, however, has no other option but to contend with its growth in the area where it is planted. As for the livestock component, it can either be allowed to graze freely in the land or it can be confined with cut-andcarry employed. This latter alternative would thus require a minimal land area.
2. *Size of the land.* The area should be big enough to supply the requirements of both crop and livestock. Because of the two production units being implemented on the same piece of land, it is expected that the normal requirement of the basic commodity may need to be expanded to accommodate the other. If cattle will be allowed to graze under the trees, the distances of coconut trees should be wider as the crops will need more sunlight to enhance their forage growth.
3. *Elevation of the land.* Climatic conditions in the slopeland areas are vital in determining the crop that can grow best as well as the livestock that can adapt to the climate in the area. Although it is true

that livestock can be provided facilities and confinement when the climate is too cold, this may be too costly for smallholders living in the slopelands. Studies show that cattle, sheep and goats can still adapt to elevation up to 1,800 meters in tropical and subtropic countries.

Labor

Since most of the inhabitants in the slopeland areas have smallholdings, the tendency is to engage in family enterprises. This situation limits the scale or size of the crop-livestock enterprise to the farmers' household. Availability of labor affects the outcome of the enterprise. The amount of labor needed depends entirely on the type of crops and animals to be integrated. Fruit trees will have different labor requirements compared to cash crops like vegetables or corn, and so with large animals versus small animals. This is further affected by the method of rearing the animals, that is, whether it be by grazing or confinement, the latter being obviously more labor-intensive.

This manual labor limitation is the reason why most of the crop-livestock enterprises in slopeland areas are smaller and often not within the economy of scale sizes. Indonesia and the Philippines have an average of 2 head of cattle per farm family, while Japan has five times more. This proves that because of improved facilities in developed countries, the carrying capacity, in terms of labor requirement, can also be increased. Therefore, in determining the size of the croplivestock enterprise, one should consider the availability of labor in the farm family. Likewise, support facilities should be considered to ensure its viability and sustainability in the slopeland areas.

Capital

Assuming that land is already available and labor will be the counterpart or equity of the farmer, additional capital is still needed for some important items, such as:

— Production inputs, e.g. seeds, fertilisers, chemicals, stocks, supplementary feeds, biologics, etc.

— Farm implements/equipment

— Production facilities like water system

— Post-harvest support facilities

— Marketing support

The slopeland areas have to utilise indigenous farming systems to reduce cost, increase profit and ensure sustainability. It is not advisable to have an enterprise which is capitalintensive in these areas.

Conservation and Management of Slopland areas

The slopeland areas which are mostly in the uplands have been considered frontier territory, a safety valve for the poor and dispossessed, and more importantly, it serves as the life support system of the lowland and aquatic areas.

Over the years, the resources present in these areas, which were once considered inexhaustible, have been expended in a prodigal manner while the population dependent upon them has grown as rapidly as ever. The ecological balance continues to decrease because these areas are now under agriculture of one type or another. This has accelerated severe ecological damage that has in turn caused the following:

— Increased runoff that has led to increasing incidence of flash floods causing severe damage and destruction to crops, properties and infrastructure in the lower region.

— Increased erosion, that has caused severe siltation, decreased water levels in rivers, and drying-up of spring sources that the populace depend upon.

— Downstream effects include a lowering of water quality through turbidity and pollution and the destruction of coral reefs.

However, if slopelands are properly managed, they can contribute greatly to arresting the social, economic, and environmental concerns of most countries. The environmental damage, particularly to the soil, is so devastating for the present generation that this should always be taken into consideration in slopeland farming.

Application of sustainable technologies

Sustainable technologies in agricultural development have become popular. It has been observed that the mainstream approach to modernising agriculture has led to dependency on external inputs such as seed, fertiliser, pesticides, machinery and fuel. In recent years, a shift away from this dependence on non-renewable natural resources was seen as imperative. Efforts have been made to promote and practice forms of land use which make efficient use of locally available resources. The concept of site-appropriate agriculture builds on a long tradition of location economics and farm management theory. Concepts being promoted are:

— Ecologically-sound agriculture

— Biological husbandry

— Organic farming

— Conservation agriculture

All refer to forms of agricultural land use which depend primarily on local resources to achieve lasting productivity. This is widely termed as sustainable agriculture with low levels of external inputs. In other countries, it is termed as ecofarming. This type of agriculture cannot be applied in an absolute sense, nor can it be expressed in quantitative terms.

Ecology is defined as a science which deals with the relationship between organisms and their environment. Thus, ecofarming or sustainable agriculture which relates to a developmental process must be continuously assessed to determine whether or not specific measures are of value. This means that farming technologies are not good or bad per se; their value depends on their applicability within a particular system.

Community Organising

A farmer may be able to implement, on his own, a crop-livestock enterprise, however small, because of the prevailing situation in slopeland areas. Data gathered showed that farmers in these areas in Mindanao have smallholdings or small farm size, limited capital, insufficient labor, difficult access to credit, poor transport system, and no postharvest facilities or marketing support. If these limitations are not arrested, the slopeland farmers are expected to remain dependent on external inputs from traders who have their own interest (profit) to serve.

To sustain the benefits gained from their individual crop-livestock enterprise or economic activities, the need for strong community or farmers' organisation is very important. It can serve as the foundation of slopeland communities' long term capability for a self-directed and self-propelled sustainable development. This in turn means less dependence on external support.

There are two main strategies of strengthening community organisations. These are the conduct of technical as well as human resource development training. These will enable them to acquire the necessary knowledge to enhance their skills in the implementation of various enterprises and economic activities as well as in managing their organisations. From these training programs will emerge leaders.

Moreover, a strong and active participation of the members in communal activities will be seen. The community's institutional development can be strengthened further through the following important elements:

- *Organisation-building* - this involves registration to acquire legal per-sonality
- *Resource build-up* - the organisation can engage in savings mobilisation as funding source of their economic activities; and
- *Linkage development* - tapping partner institution for common benefit of the members of the organisation.

Having advanced the status of the organisation, the benefits of community organisation in a croplivestock enterprise in slopeland areas include:

- Easy access to credit
- Bulk procurement of feeds, fertilisers, etc. that can reduce cost
- Complementation of feeding between members
- Ease and profitability of group marketing of products
- Group-facilitated veterinary and animal health care services
- Establishment of communal rather than individual support facilities that can save cost.

— Easy and cheaper investment for establishing communal enterprises for the members.

The success indicators of community organisations can include: diversification and/or expansion of viable enterprises, consolidation of resources, savings or capital build-up generation, creation of umbrella organisations or federations, realisation of marketing tie-ups with businesses, and stronger working linkages with other institutions.

Socio-economic Indicators

Land ownership/tenure status

The type of crops to be planted depends greatly again on the tenure status of the land to be developed. Permanent crops like fruit trees which have 5 - 50 years of economic life span will be most likely to be implemented only in lands with ownership or stewardship. In the case of animals, they can be transferred from one place to another with no direct relation to land ownership or tenure status, as the feeding system can easily be shifted.

In the Philippines, lands are classified into two types, Public Forest Land for which no ownership titles can be issued; and Alienable and Disposable (A & D) lands for which can be issued permanent ownership titles. Classified as Public Forest Lands are those with 18% slopes and above. They can still be developed with the issuance of stewardship certificates for 25 years, which are renewable for another 25 years. Those below 18% slopes are A & D which can be developed by the owner with no reservations.

In Mindanao, it has been proven that where land tenure is not a problem, development of crop-livestock enterprises is faster and more successful. This is because farmers are not hesitant to invest in land which they know will not be taken from them. Permanent crops like fruit trees and coconuts are usually integrated with animals. Successful crop-livestock enterprises are thus mostly found in areas below 18% slope or above with stewardship agreement. The thinking of many farmers is that permanent crops planted will help in securing ownership or stewardship of public lands.

Role of farmers' organisations

Farmers organisations serve as the nerve center of communal activities in slopeland areas. These organisations provide a venue to discuss problems and solutions related to their enterprises. They also facilitate active cooperation among the members, enabling them to address the higher needs of of their produce, accessing credit and others. In other words, the production stage of the enterprise can be assisted by the farmerorganisation in many ways.

One important role it can play as well is on policy advocacy matters related to crops and livestock. For crops, the organisation can lobby, for instance for the reduction of prices of production input, and support for the improvement of facilities like farm-to- market roads which normally could not be provided to one farmer alone. For livestock, the farmers can express their sentiments on importation of meat and other livestock products. Through the farmer's organisations, the government can be influenced to formulate and implement policies that will protect the interest of the crops and livestock entrepreneurs.

Labor distribution and manpower needs

In a mango-goat enterprise, where the area of the land utilised is ½ ha and the number of goats is 1 doe and 1 buck for one farmer, labor distribution at one production cycle is 90 full man-days for mango and 15 full man-days for goat. Mango production will need at least one person to work on land preparation, planting, and tree guard installation for 30 man-days.

After that, maintenance work will need a part time person up to harvesting for 60 man-days. The goats, on the other hand, will need only 14 man-days from forage establishment to main-tenance. In this enterprise, the main commodity is mango; therefore, the farmer should concentrate on it while the goat can be taken care of by other members of his family. In this situation, the goat can be confined, and the use of a cut and carry system is appropriate. One important thing to consider therefore, is to select the right crop at the right time to be integrated with the livestock and viceversa. If one commodity is laborintensive then the other must be less labor-intensive to avoid problems of providing labor requirements of two commodities at the same time.

Emerging role of women and children

While the farmer (normally the husband) works full time in the farm, the wife and the children can provide the labor requirements of the other enterprises within the farm. This is especially useful if there are two production units or commodities on the same piece of land. The wife and children do the lighter load of weeding, applying fertilisers, watering, planting, harvesting and maintenance. As much as possible, they do the work that can be delegated by the farmer husband so that their enterprise can be expanded.

Children, however, have limitations on time as they are attending classes in school. In underdeveloped countries, however, they are expected to work in the farm on off days from school. There are still a lot of school children who are forced to stop schooling and are obliged to work in the farm to earn a living. This is due to the widespread poverty in slopeland areas.

Participatory communication

Participative community based approach is the key to sustainability of development initiatives. This also applies to crop-livestock enterprises where farmers should be involved from the planning and implementation to the evaluation stages. It is important that the beneficiaries of any development programs, the community and the implementors, jointly appreciate the relevant information needed in determining the enterprise to be established.

Based on SMAP development process, the following stages of developing an enterprise necessitates full participation/ involvement of the farmer beneficiaries together with their families and the community officials or leaders:

Stage 1. Data collection. Before identifying an enterprise, it is very important to understand the community situation, which needs development of a community profile with the following information:

1. People
   - The number of men, women and children (by age group)
   - Social groups (ethnicity, language groupings)
   - Household characteristics (size and location, length of stay in the community)
   - Number of people by occupation and education level
   - Structure of society (affiliation to churches and membership in organisations), leadership

2. Resources
   - Community natural resources (water sources, fuelwood lots, etc.)
   - Land classification and use (alienable and disposable, forest, etc.)
3. Facilities
   - Community facilities (churches, schools)
   - Health clinics
   - Water supplies
   - Communications (roads, vehicles)
   - Multi-purpose pavement
   - Barangay Hall, Multi-purpose building
   - Power (electricity)
4. Seasonality
   - Rainfall
   - Crop/harvest
   - Labor in agriculture (periods of much and little work)
   - Food availability (shortage, abundance, type)
   - Illness by type (for people and animals)
   - Fiestas
   - Water supply
   - Fuel availability
   - Income/credit (times of need, for what)
5. Farms
   - Size
   - Land tenure
   - Soil type

— Cropping pattern (temporary crops)
— Permanent crops
— Livestock
— Inland fisheries

6. Markets
   — Source of farm inputs
   — Farm product outlet
   — Marketing organisations; and
   — Buyers organisations

Stage 2. Identification of Problems, Needs, and Solutions. It is important as well for the beneficiary and the community as a whole to agree on the priority problems and needs to be addressed, and also the solutions. Capability to plan and implement the identified solutions through projects or enterprise establishment should be considered.

Stage 3. Preparation of the Feasibility Study. The enterprise or project identified must have a study to determine whether it is technically, financially, socially and environmentally feasible. Data must be validated jointly by the community and the intended beneficiaries or farmers.

Stage 4. Implementation Stage. There should always be discussion by the farmer with the community on the problems affecting the implementation. The farmer may need training to enhance his skills. Thus, it is important to identify who needs to be approached for support in the community before considering outside support.

Stage 5. Monitoring and Evaluation. Farmers should know if the enterprise is performing according to expectation. They must ensure that there are no violations of laws or negative impact to the populace. After evaluation, the farmer and the community may inject improvements on the technologies to increase productivity and profitability of the enterprise.

Role of cultural values and traditions

In most cases, the cultural values and traditions influence the motivation which results in the active participation of the people in the community. If the enterprise is parallel to their cultural practices, there will be no problem of accepting responsibility by the people concerned which is of critical importance to the success of the enterprise. For example, a pig project will not be acceptable to the Muslim community.

Decision-making has to take into account the existing system in the community. Under the Muslim community again, the Datu or Head should agree to the activity or enterprise, otherwise the support of the clan will not be to the fullest. Involvement of women in the implementation, and their sharing of the fruits of the enterprise as well as access to factors of production and support services, and their rights such as land ownership and inheritance, must be considered.

Economic Indicators

Capital investments

The investment needed in a crop-livestock enterprise could be sourced from different entities through:

— the farmer himself

— loans

— partnership

— savings

The capital requirement need not necessarily be contributed all at the same time. For example, surplus cash earned by farmers is often invested first in livestock and the offspring of the animals provide the income. Animals in turn can be used to finance cropping inputs. This can also be done in another way, in which investment goes first to fast-income-generating crops or cash crops to enable the farmers to accumulate capital to invest in livestock. In this respect, the integration should be properly timed. It can be scheduled by phases especially if there is limited capital. Investment planning is therefore the key to optimise the productivity of the capital intended for any enterprise.

Economic returns from the integrated system

In terms of economic viability, the enterprise returnon-investment should not be below the prevailing stock market rate or prime interest of the banks. From the economists' point of view, the cost:benefit ratio can also serve as the indicator of viability, such that when the ratio is high, income exceeds cost. If the enterprise was funded by a loan, economic viability could also mean that payback period is shorter than the economic life span of the enterprise. In computing the economic returns for the integrated system, it is important to include the by-products of the main crops as well as the value of the inputs resulting from the interaction of the two production units. As in the case of the mango-goat enterprise the following are some of the additional benefits derived that added value to the enterprise:

- Manure of the goat
- Nitrogen fixing benefit of legume trees such as Renzonii and Flemingia restoring fertility to the soil
- Cash crop production while mangoes are still small
- Services of the buck which are paid for by other goat farmers

Need for markets to absorb the produce

No enterprise will succeed without markets. In SMAP, marketing is viewed as a planning process of enterprise development which does not only cover selling. It is a process by which the stages of enterprise development are related to marketing such as:

— *Identification of crops/livestock.* Select only those that are marketable or in demand.

— *Timing of production.* Aside from seasonality, harvesting should be timed to high demand that falls on lean months. It is also important that drying, storage, and transport of the product to the market be facilitated.

— *Processing.* The potential of the product for processing to have added value is a major consideration in selecting the crop. It provides opportunity for high profit and diversified marketing for the produce to be all absorbed in the market.

4. *The need for credit to support small farmers with little capital.* Credit is a factor that can sustain the enterprise or even diversify/expand it if the resources and market will allow. Slopeland farmers though have problems in securing credit for of various reasons:

   — Land tenure problem

   — Lack of collateral

   — Stringent requirements of banks

   — Need for groups/organisations to be registered

   — Insufficient capital or equity

   — Insurance scheme needed to cover crops/ livestock in the slopelands

Without credit facility or access to financing, the small farmers will continue to depend on traders who charge very high interest rates and buy farmers' products at very low prices. Access to credit will result in less dependency on traders and more profit for the enterprise.

Analysis

The socioeconomic considerations and indicators discussed will be very important and meaningful for the farmer-entrepreneur who plans to engage in any crop-livestock enterprise. It is equally important to understand that the two aspects social and economic may not necessarily meet, and many times one is sacrificed because of the hierarchy of needs in a given community. Sometimes the enterprise is economically viable and an income earner, but it is not socially acceptable as it is against the norms or cultural practices of the people in the community. More importantly, the other aspects like environmental concerns and sustainability must be given equal consideration and attention to ensure success of the enterprise.

Policy Concerns and Issues

There are only few concerns and issues that need to be attended to in the crop-livestock enterprise in the slopeland areas. These are:

— *Land tenure*. The government should come-up with measures to improve the distribution and ownership of the land to optimise the utilisation of this resource.

— *Conservation and protection laws* formulated and implemented. This is the only way that upland resources can be preserved for the next generation.

— *Disparity of support* by all sectors between slopeland and lowland farmers must be reduced.

— *Education or value orientation* will help reduce ecological problem caused by lack of knowledge on environmental issues.

The integration of crop-livestock in slopeland areas is a common practice in many countries. The most important thing that needs to be ensured is the sustainability of the crop-livestock enterprise. It can be summarised, therefore, that the following factors should be considered to ensure sustainability of the enterprises in the slopelands:

1. Policy support
2. Appropriate technology
3. Environmental protection
4. Socio-cultural aspects
5. Enhanced institutional and management capacity
6. Economic and financial analysis showing high viability.

The government and private sector are therefore encouraged to give their sincere contributions and concerns in ensuring that the above factors are met and made relevant to the agricultural activities in the slopeland areas.

8

Mixed Farming and Environment

Mixed farming systems, the largest category of livestock system in the world, cover about 2.5 billion hectares of land, of which 1.1 billion hectares are arable rainfed crop land, 0.2 billion hectares are irrigated crop land and 1.2 billion hectares are grassland. Mixed farming systems produce 92% of the world's milk supply, all buffalo meat and approximately 70% of the sheep and goat meat. About half of the meat and milk produced in this system is produced in the OECD, Eastern Europe and the CIS, and the remainder comes from the developing world. Over the last decade, meat production from this system grew at a rate of about 2 percent per year and thus remains below global in demand.

Mixed farming is probably the most benign agricultural production system from an environmental perspective because it is, at least partially, a closed system. The waste products of one enterprise, which would otherwise be loaded on to the natural resource base, are used by the other enterprise, which returns its own waste products (manure) back to the first enterprise. Because it provides many opportunities for recycling and organic farming and for a varied, more attractive landscape, mixed farming is the favourite system of many agriculturalists and environmentalists.

In many situations crop and livestock production is largely in balance with nature. There are important exceptions, such as some mixed farming systems of the tropical highlands of Asia and Central Africa which, partly because of overgrazing, are amongst the most eroded and degraded systems of the world. On the other end of the development spectrum, heavy use of feed and fertiliser in the industrial world and in some of the fast growing economies of East Asia, has led to nutrient loading, habitat destruction and water pollution. In this context, it has to be remembered that integrating crops and livestock neither generates new nutrients nor reduces nutrient surpluses.

Evolution of Mixed Farming System

As rural population pressure increases, both crop and livestock farmers need to intensify production. McIntire *et al.*, show that, as population pressure increases, the two activities often become integrated. Traditional soil protection techniques, in particular the long fallow periods which protected against erosion and allowed soil nutrients to recharge, are no longer possible. If farmers cannot resort to external inputs, the integration of livestock and crop activities represents their main opportunity for intensification. Mixed farming has therefore become the basis for modern agriculture.

Mixed farming systems provide farmers with an opportunity to diversify risk from single crop production, to use labour more efficiently, to have a source of cash for purchasing farm inputs and to add value to crops or crop by-products. Combining crops and livestock also has the potential to maintain ecosystem function and health and help prevent agricultural systems from becoming too brittle, or over connected, by promoting greater biodiversity, and therefore increased capability to absorb shocks to the natural resource base.

Environmentally, mixed farming systems:

— maintain soil fertility by recycling soil nutrients and allowing the introduction and use of rotations between various crops and forage legumes and trees, or for land to remain fallow and grasses and shrubs to become reestablished;

— maintain soil biodiversity, minimise soil erosion, help to conserve water and provide suitable habitats for birds;

— make the best use of crop residues. When they are not used as feed, stalks may be incorporated directly into the soil, where, for some time, they act as a nitrogen trap, exacerbating deficiencies. In the tropical semi-arid areas, termite action results in loss of nutrients before the next cropping season. Burning, the other alternative, increases carbon dioxide emissions; and

— allow intensified farming, with less dependence on natural resources and preserving more biodiversity than would be the case if food demands were to be met by crop and livestock activities undertaken in isolation.

Under different sets of pressures and opportunities, several developments are possible, depending on resource endowment and market access. Initially, and if market opportunities open up, the symbiosis between crops and livestock can intensify, and the nutrient balance can be maintained.

If pressure increases, crop-livestock systems can separate into specialised crop or livestock activities. If there are no improved market opportunities, which is the case in many developing countries, and as human population pressures continue, the arable land part of the system will experience increased rates of nutrient depletion and soil erosion. This can, in turn, lead to a downward spiral of mono-culture with lower quality food crops, increased under-nutrition and famine.

However, if urban incomes rise, more market opportunities open up and farmers become more integrated into the market economy, allowing them to specialise, take advantage of economies of scale and develop greater levels of expertise. Finally, under very strong demand, and often encouraged by input or price subsidies such as exist in many developed countries and in the fast growing economies of East Asia, excessive importation of nutrients can lead to soil and water tables being overloaded with nitrogen or phosphorus.

The challenge for the mixed farm sector will therefore be to maintain a nutrient and energy equilibrium through crop-livestock integration and at the same time allow sustainable productivity growth. The mixed farming system, more than any other production system, operates under a wide range of environmental and economic conditions and requires regional solutions and practices.

However, there is one overriding criteria in determining the size and nature of the system's impact on the environment, and this is the nutrient balance. This balance is determined by the nutrients (N. P and K) brought into the farming systems by inorganic fertiliser, feed, nitrogen fixed by leguminous plants and transfer from grasing areas outside the farm, and the amounts exported in animal products or lost from the land to the air or groundwater. A positive balance of nutrients will have a completely different effect than a nutrient deficient system. In this analysis, the mixed farming systems are classified either as nutrient deficient systems, which occur mainly in the developing world, or as nutrient surplus systems, mostly found in the industrialised world and increasingly in the fast growing economies of East Asia.

Nutrient Deficits in Mixed Farming Systems

The mixed farming systems of the developing world contain about 67 percent of the cattle and 64 percent of the small ruminants of the world. Throughout the world, animal numbers are growing in the mixed farming system, most rapidly in the humid/sub-humid regions. Sheep and goat numbers show the fastest growth rates in the humid/sub-humid region, underlining how human population pressure is reducing farm size and access to and use of resources.

Irrigated mixed farming systems have shown the greatest increase in productivity, particularly in the humid regions of Asia. This is clearly a result of the strong growth in demand for animal products and better access to feed resources and other types of infrastructure in Asia. Milk production is important, particularly in south Asian countries, due to the growing demand in the region and the favorable policies that many governments have created for the dairy sector. Although the growth of dairy production can place more pressure on land resources, it can also increase the use of crop by-products which in turn improves nutrient recycling and, if of high quality, can diminish methane production.

There is a considerable range of positive and negative interactions between livestock and the environment within the different sub-systems. These interactions can affect land quality in its physical (soil erosion) and chemical properties (soil fertility), the use of nonrenewable resources, such as fossil fuels and fertiliser, and the conservation of agricultural (plant and animal) biodiversity. They are detailed below.

Soil erosion is probably the most pervasive form of land degradation in the developing world. Erosion rates are particular high in Asia, Africa and South America where they average 30 to 40 tons/ha/year, compared with an average soil formation of 1 ton per ha/per year. In Africa, 60 percent of soil erosion damage occurs in the semi-arid and dry sub-humid regions where cropping and livestock co-exist, with strong interlocking factors of cropping, fuelwood collection and grazing.

Soil erosion is even more damaging on sloping lands. Poorly managed sloping terraces under crops and degraded rangelands can lose up to 100 MT/ha/year. Conversely, well managed pasture land loses 7 MT/ha/year or less and well managed forest land losses range from 0-10 MT/ha/year. Bojo and Cassells quantified these losses in the degraded Ethiopian Highlands. Their results showed the effect of including pastures in those environments as a means of reducing erosion.

Soil Fertility

Livestock play a significant role in maintaining soil fertility. In partially closed mixed farming systems, livestock can replenish a substantial share of soil nutrients, and therefore reduce the need for inorganic fertiliser, with corresponding savings for farmers in terms of cash outlay, for the country in foreign exchange, and for the world in non-renewable resources. The total value of this contribution is not known, although approximate figures are given by Jansen and de Wit for the irrigated mixed farming systems of Asia. Assuming manure production of about 1 ton dry material per year per Tropical Livestock Unit, and an effective availability of 15 percent of the nutrients in the case of stall fed and half that amount for grazing animals, the amount of nutrients under stall fed conditions is about 8 kg N and 6 kg P per year per TLU. Under those assumptions, livestock in the mixed irrigated farming system can supply between 2 and 10 percent of the nitrogen requirements for rice, and about 40 to 120 percent of the requirements for phosphorus in cassava.

Nitrogen is the most depleted nutrient, with the amount lost varying, depending upon the technology used in collection, storing and application practices. How farmers approach these three factors largely determines the effectiveness of the recycling process and therefore provides an avenue for intervention. As shown, stall feeding significantly increases the amount of nutrients available from manure.

To demonstrate the importance of these contributions, an estimate was made of the amount of fertiliser that would be required to replace the manure that is used in the irrigated system of the humid tropics. The value of this is estimated to be between US$ 700 to US$ 850 million per year, depending on various assumptions.

The economic benefits of improved soil structure as a result of adding soil organic matter from manure are more difficult to estimate. Adding manure to the soil increases cation exchange capacity, and improves soil physical conditions by increasing the water-holding capacity and improving soil structure stability. When adding the manure output from pigs and ruminants together, livestock may contribute up to 35% of the soil organic matter requirements. This is a crucial contribution because this is the only avenue available to many farmers for improving soil organic matter.

Non-renewable Resources

Draught animal power, in addition to other economic benefits, provides an important source of energy in the mixed farming systems of the world and helps to reduce dependence on nonrenewable fuel resources. In developing countries, animal power is used to cultivate about 52 percent of the 480 million hectares of cropland. Draught animals provide between 25 to 64 percent of the energy needed for cultivation in the irrigated systems of the world.

Worldwide 300 million draught animals are used in small-scale agriculture, while 30 million tractors would be needed to make the same contribution. This is equivalent to an US$ 200-300 billion investment in tractors plus a $5 billion annual fuel cost Assuming that approximately 50 percent of the arable land is cultivated in the world's irrigated areas with draught animal power, complete replacement by tractors would require annually 888 million litres of diesel at a cost of US$ 222 million, and US$ 429 million for depreciation of the equipment.

The value of animal manure for cooking fuel is more controversial. Dung cakes are the main source of household energy for millions of poor households in the developing world. It is often argued that using dung for fuel removes soil nutrients and therefore carries high opportunity costs. However nitrogen is the only major nutrient that is lost because the remaining ash is high in phosphorus and potassium.

Biodiversity

In mixed farming systems, livestock-environment interactions can be a key focal point. As human population pressures increase, some types of wildlife or plant diversity can be lost while other types of biodiversity may actually increase. Reid et al. demonstrate how the number of tree, bird and mammal species changes as human use of the resource base intensifies. These results are important for demonstrating the dynamic nature of ecological systems. They also show that man's agricultural activities must be balanced with the environment's capacity to sustain them.

An often overlooked component of biodiversity is soil micro-flora and fauna. Arthropods, with 90 percent of all species, dominate biodiversity. For example, in a New York alfalfa "ecosystem" Pimental et al., reported 600 species of above ground arthropods. Livestock and manure have a beneficial effect on this biodiversity. For example, in mixed farms in Japan, the species diversity of the micro-fauna under grass more than doubled when manure was added to the land.

In forest and communal grazing areas, adjacent to mixed farming areas, plant and animal biodiversity is decreasing because of over-grazing. There are many examples quoted in the literature confirming this trend, for example, in Syria, Rajastan, and West Africa. It is not clear, though, whether these are irreversible, long term changes. For example, Thomas and Middleton report that no systematic pattern of vegetation change could be detected in a 27 year vegetation survey of 77 villages in the Sudan. However, wildlife biodiversity is disturbed if land is fragmented and river sides are cultivated. Finally, the mixed farming areas of the world contain the main centres of domestic animal genetic resources of the world.

Driving Forces

Land. Human population pressure, poverty and infrastructure are, as in grazing systems, the most important fundamental driving forces affecting the environmental impact of mixed farming systems in the developing world. And in that respect, the mixed farming systems are entering a dynamic period of growth and change because they cover some of the countries with the highest birth rates in the world. Human population pressure affects soil fertility. In many parts of Africa and in the mountainous areas of Asia and Latin America, farmers attempt to maintain soil nutrient balance with, at best, small inputs from outside.

Poverty and poor infrastructure usually prevent them from buying commercial feed and fertiliser. Nutrient balance depends, therefore, on the ratio between arable and non-arable land. This also varies widely according to the climatic conditions. For example, in West Africa, the ratio was about 10 to 40 hectares of dry season grazing and 3 to 10 hectares of wet season grazing to provide adequate nutrients for one hectare of millet cropland.

However, with increasing population pressure more and more grazing land is converted into cropland, the crop/grazing land ratio narrows, and the nutrient flow into cropland decreases. For example, cropland has increased in the Machakos area of Kenya from 35 percent of the higher potential area in 1948 to 81 percent. In Northern Mali, the rate of increase of the arable cropping area was the same as the 2.6 percent rate of annual population growth over the last twenty years. This trend can lead to increased competition for cropland and grazing on the open rangelands, which in turn, can lead to privatisation of crop residues and rangelands.

For example, the common property lands in India, have decreased since 1960 by 50 percent, so that in 1992 they constituted only 15 percent of the total area. Population pressure also causes farmers to push cultivation out into more resource limited areas, causing soil erosion. Mabbutt reports, for example, that in Niger millet fields are appearing 100 km north of the recommended limit to cultivation, increasing the risk of crop failure and leaving soil exposed and at risk of erosion. Livestock are moved to the most marginal areas.

Unless nutrients are replenished from outside sources, soil fertility continues to decline as the ratio between crop and grazing land declines. This is typically the case of many mixed farming systems in the tropics. Reported losses range from about 15 kg N/ha/year in Mali, to more than 100 kg N/ha/year in the highlands of Ethiopia.

Downward spirals of declining soil fertility, overgrazing, and increased erosion and losses in soil micro-flora and fauna are the result and will cause, ultimately, significant decreases in agricultural productivity and political stability. The Rwanda example clearly underscores the need for a balance between crop activities, animal inputs and human population pressure, to avoid the collapse of the farming systems, impoverishment and even civil war.

In irrigated mixed farming, grazing is more likely to be contained within the system although the source of feed supply is likely to shift increasingly from non-arable. grazing to grazing of crop residues and the. use of fallow land. Inorganic fertilisers are used to overcome soil nutrient deficiencies.

Policies have often limited the beneficial impact of crop-livestock integration and exacerbated its negative environmental effects. These policies have been instigate in some countries by a desire to achieve self-sufficiency, especially in cereals, and in others by the desire to provide cheap food for urban areas. To achieve self-sufficiency, markets were protected and subsidies were provided on inputs such as fertiliser, feed, fuel, etc. in order to increase production. This created disincentives for the use of on-farm products such as crop-residues, animal draught power and manure and contributed to a decline in mixed farming, with subsequent negative effects on the environment. More specifically:

— subsidising fertilisers, common in the oil-producing countries such as Nigeria, and many of the North African and Middle Eastern countries, hindered the integration of crops and livestock because it reduced the need for organic fertiliser;

— subsidising mechanisation and fuel, again common in the Middle East, North Africa and Asia, reduced the need for animal traction and, as mentioned before, made possible the expansion of arable farming into marginal areas;

— subsidising concentrate feed, common in the Middle East and, indirectly through differential import duties on feeds in the KU, leads to the excessive use of concentrate feeds. It also reduces the need for mixed farming. Globally, it contributes to a "feed nutrient trade deficit" in the developing world; and

— poor land tenure security, especially in the rainfed mixed farming systems of the developing world, has discouraged investment in long-term soil fertility improvements, such as the use of inorganic fertilisers and the use of green manure and leguminous fodder crops within the crop rotation.

The policy of providing cheap food to urban centres has at the same time discouraged intensification, in some cases to the detriment of agricultural production as a whole and livestock in particular:

— imposing high import duties to protect domestic cereal production, has had the effect of extending cropping into marginal areas and upsetting the equilibrium between crops and livestock;

— overvalued exchange rates in sub-Saharan Africa and Latin America have favoured imports of cheap food from the industrialised world, thus competing against local production and providing no incentives for local producers to intensify into mixed crop/livestock systems nor to practice soil conservation.

Technology and Policy Options

A fundamental requirement to achieve environmental sustainability of the natural resource base is to accept change and flexibility in the respective production systems and overall landscape. Agriculture, and crop agriculture in particular, seeks to stabilise production and limit environmental change thereby reducing the flexibility of the eco-system. But any ecosystem which supports the production of food, can be enhanced by the integration of livestock as a mechanism to promote system flexibility.

Mixed farming, with careful attention to the nutrient balance, is therefore probably the most environmentally desirable system, and should be the prime focus of agricultural planners and decision makers. The challenge will be to identify the technologies and policies, which generate the sustained and accelerated growth needed to satisfy the world's booming demand for meat and milk. This challenge will be enormous.

Past global growth of the meat and milk output of the mixed system has remained behind the increase in global demand. Will the growth in demand levels continue to rise beyond the capacity of mixed farm systems to meet national demands and therefore result in an immediate transition to highly intensive industrial systems? This study argues that, for environmental reasons, it is important to promote mixed farming in order to satisfy, as far as possible, global demand for meat and milk.

The key enabling factor seems to be access to inputs and attractive markets, as illustrated by the Machakos case study in Kenya. Here, a dynamic market and access to inputs has led to a fully sustainable development of a resource poor area, in which the population had quintupled over the last six decades. Use of livestock in such intensive situations includes a shift in species from cattle to sheep and/or goats.

In such intensive situations, sheep and goats are kept where difficult access to markets and milk processing facilities make keeping cattle a less attractive option. Integrated crop-livestock systems have an immense potential to contribute to the required growth in productivity in the highland areas of Latin America and the semi-arid areas and highlands of sub-Saharan Africa where the population density is already high.

In spite of all the advantages of mixed farming, current trends point to specialisation in either crop or livestock production, whereby the rate of change in any particular situation will depend on existing infrastructure, relative price ratios between the different inputs and outputs, and economies of scale. The development path that has been followed in the developed world and east Asia confirms this trend.

In mixed farming systems, there are exciting opportunities for technological change. Applied research and extension are of critical importance if the environmentally friendly factors of the system are to be maintained, although McIntire et al., make the point that African farmers are well aware of the technologies involved. The challenge will be to convince farmers about the value of technology. In the nutrient deficient systems of the developing world, the emphasis should be on control of soil erosion and improving nutrient recycling. Some examples are:

- Improvement of soil cover through the use of alternative crops for mulching, and introduction soil management techniques such as conservation tillage, bench terracing, strip cropping, contour farming, etc.;
- Improvement of feed production and quality to reduce the pressure on grazing areas and improve internal nutrient transfers. Technologies to do so include:
  - introduction of fodder shrubs and trees to reduce soil erosion and improve soil fertility. Several mixed farming systems using fodder shrubs and trees have been developed. An example is the agroforestry system with three strata (grass, fodder shrubs, and tree crops, such as oil and coconut palms, cashew nuts, etc.) as successfully introduced in Indonesia;
  - improvement in feed quality, for example through urea treatment of feeds, as reportedly successfully used now by more than 5 million farmers in China and through increased efforts in plant breeding to correct the grain bias of the green revolution "dwarf' varieties and raise nutrient quality of the crop residues; and
  - use of non-conventional feeds.
- Reduction of nutrient losses from manure and improved efficiency of their application by:
  - promotion of stall feeding which doubles the effective availability of nitrogen and phosphorus; and
  - strategic supplementation for specific classes of animals to improve the efficiency of limited amounts of available feed.

— Increased production efficiency, and thereby farm income, resulting in improved purchasing power for soil improvement and conservation methods. They include:

  — within breed improvement;

  — where appropriate, changing to small ruminant production or to optimal combination of ruminant and non-ruminant species; and

  — strategic supplementation of lactating and growing animals.

Clearly, for optimal development of mixed farm systems, a more open market economy is needed. In an open market there should be reduced subsidies for feed, fertiliser and mechanisation. Phasing out of subsidies will promote a closer integration of crop and livestock systems in many parts of the world. It will enhance the use of home grown feeds, organic fertiliser and animal traction. To capitalise on free market transitions, better extension, improved financial institutions, security of tenure and, above all, better infrastructure are required.

The greater reliance on market forces needs to be accompanied by policies which seek to provide, for the most densely populated areas, the highest priority for employment generation outside the sector. This can be accompanied by market development for intensive crop and livestock production, such as horticulture and, in livestock, in industrial pig and poultry and possibly small ruminant production. This will enable adequate flows of nutrients in to the system and a rebuilding of environmental sustainability.

For the less critical, but still densely populated areas, markets and infrastructure are still important. For those systems, closer crop-livestock integration will help to reduce eventual nutrient deficits, but markets will always be required to supply external inputs. Technology and market policies should encourage rotations that maintain soil fertility and find synergies between crop and livestock activities.

Mixed Farming and Nutrient Surpluses

The mixed farming systems of the developed world cover about 17 percent of the world's pasture land and half of its arable land. They contain about one-fifth of both the world's cattle and small ruminant population. These systems generally show a much slower growth than mixed farming systems in the developing world. In some countries, mixed farming is contracting, reflecting the combination of stabilised demand in the OECD countries and the decline in the former Soviet Union and Eastern Europe, because of restructuring following the transition to a market economy.

With increasing population pressure, growing incomes and improved infrastructure and market opportunities, more intensive forms of crop and livestock production, including integrated systems evolve. These integrated systems come into dis-equilibrium in several regions of the world as a consequence of large nutrient imports from. outside the region, causing an overloading of soil and water with pollutants. The environmental challenge in the industrial countries, and to a large extent also in the fast growing economies of east Asia, is thus to identify, before it is too late, the point where land and aquatic ecosystems become overloaded. This is better than trying to identify technologies and policies which mitigate the negative effects of nutrient overloads once they have occured. In other words, "prevention is better than cure".

A further environmental challenge is to maintain biodiversity and an aesthetic landscape in heavily populated and developed mixed farming areas. Although opportunities to conserve biodiversity will not be able to mimic exactly the diversity associated with mixed farming in less developed areas, it should be possible to achieve some level of conservation. The challenge is to put in place technologies and policies which encourage farmers to retain a mixed farming system instead of specialising in the production of one or two crops only.

In the developed world, the environmental effects of mixed farming systems may include a deterioration in land and water quality and biodiversity loss, especially through habitat change. Soil erosion in the temperate zones, with a loss rate of about 15 ton/ha/yr, is half that of the developing world. But even this lower erosion rate exceeds by far the average soil formation of about 1 ton/ha/yr. Soils in north western Europe, in the eastern and midwestern USA and in the fertile and densely populated, and increasingly affluent, areas of east and south Asia, often reveal a surplus of nutrients. There are also widespread areas of central and eastern Europe where, as a result of very large livestock production units, there is a serious surplus of soil nutrients. While the total area with surplus nutrients is probably still very small less than one percent of global arable lands, through runoff its effect on water quality is much more widespread. Often such areas are near ports and/or large urban areas so that the transport costs associated with imported fertilisers, and the subsequent sale of milk and meat, are low.

The excess nitrogen and phosphorus "leaks" through leaching or run-off in surface or groundwater, damage aquatic and land eco systems. In Pennsylvania about 40 percent of the soil samples taken from dairy-crop farms revealed excessive phosphorus and potassium levels. Soils are saturated and surplus nutrients leach into surface water, pollute the environment, in this case the Chesapeake Bay. A similar situation is found in Brittany, France where all eight counties report nitrate levels of more than 40 mg/litre whereas, in the 1980s, only one county reported levels this high.

Biodiversity. The intensification of mixed farming and the subsequent shift in production towards regional specialisation puts pressure on plant and animal biodiversity. In addition, the move towards sown pastures coupled with high levels of organic and inorganic fertiliser use reduces the richness of flora and fauna. This problem can be exacerbated by the use of insecticides for the control of external parasites, especially in areas with high water tables.

Driving Forces

Soil erosion is strongly influenced by grazing pressure and cropping intensity. Reynolds *et al.*, reported how variable soil erosion can be observed under different grazing pressures. Soil loss on lands with a good grass ground over has been estimated at 1 ton per hectare, this increased dramatically on overgrazed pastures to a level of 53 tons per hectare. If grazing pressure is in balance with the forage resource being produced, livestock's interaction can encourage more stable land use practices. Pimental *et al.*, discussed how, in the United States, in order to increase farm size, grass strips and shelter belts were removed, thus increasing the erosion rate. In the past, livestock provided a rationale for grass strips and shelter belts but, with the separation of crop and livestock systems in OECD countries, the rationale for using land resources in this way has weakened.

Nutrient surpluses are the main cause of deteriorating land and water quality. These surpluses come from a combination of inorganic and organic fertilisers. Inorganic fertiliser is used, in spite of the fact that the system could be balanced using the nutrients supplied by the manure produced within the farming system. For example, in Brittany, organic manure produced by intensive industrial and mixed farms of the region could provide adequate nutrients but substantial surpluses emerge because of the additional import of large quantities of inorganic fertilisers. Inorganic fertiliser is said to be necessary because of the ready availability of its nutrients and its easy transport but, as in pure intensive grazing systems, lower inputs might be adequate and even more economical. The preparation of nutrient balances on a regional basis is therefore a critical element in identifying land conservation and fertiliser policies.

Policies designed to protect domestic production of most animal products, especially beef and milk, against imports and a broad array of producer subsidies, especially on feed imports, have led to the wide use of feed concentrates and therefore to situations where nutrients are in surplus. Such situations occur most often on the many small-scale enterprises which tax advantages and quota systems usually favour and where regulations are more difficult to enforce. Smaller size enterprises tend to have more lenient standards and are able more easily to exploit loopholes in regulations designed to protect the environment. These intensive systems using concentrate feed demonstrate how policies designed for social objectives can misdirect technology development.

Societal Response

In the developed world, regulations are being introduced to restrict the emission of nutrients in the case of point source pollution, and restrict stocking rate in the case of non-point source polluters. A variety of measures may be introduced to reduce nutrient surpluses. These include:

— taxation of inorganic fertiliser, in order to discourage unnecessary use where a system is already in surplus;

— maximum application limits and regulation of times of application in order to reduce leaching and volatilisation;

— taxation on feed imports to reduce the significant transfer of nutrients from the already nutrient deficient developing world to the nutrient surplus areas of the industrialised world. Such taxation might be supplemented with, or used as, subsidies for a transfer of nutrients back to the developing world;

— incentives to achieve a more balanced distribution of crop and livestock activities.

Of particular importance are the following technologies to reduce nutrient surpluses:

— Improved nutrient formulation in terms of incorporating optimum levels of nitrogen into feed and adding enzymes to improve the utilisation rate of plant phosphorus. Optimum nutrient ratio and composition management reduce loading the environment with these elements; and

— Improved manure handling and storage are key requirements for more efficient use in nutrient deficient areas, and to prevent leaching and pollution of surface water where there is a nutrient surplus.

In addition, over the last decade, there has been considerable interest in promoting low-input mixed farming systems, as sustainable and environmentally friendly systems. In the USA, the Rodale Institute in Pennsylvania, and the LISA (Low Input Sustainable Agriculture) movement, strongly promoted by USDA have been in the forefront. In Europe, the ILEIA (Institute for Low External Input Agriculture) has been one of the prime movers. Mixed farming systems are often especially suitable for low input production. Some of the main technologies which can be used in the livestock sector are:

- use of grass-legume mixtures as the basic feeding system for ruminants. The nitrogen fixation of legumes (such as clovers), can replace the nitrogen lost through milk and meat;
- mixed species grazing (cattle and small ruminants), which increases grass yield, botanic composition and reduces the parasite load;
- genetic improvement focusing on maximum capacity to digest roughage, rather than on concentrate feed consumption, and based on life-time production, rather than on short term lactation yield, to achieve optimum efficiency on the basis of renewable resources;
- housing, which allows adequate movement and is based in the animals' natural social behaviour, produces high quality manure using large quantities of straw;
- disease control based on good feeding and breeding practices, and preferably using organic medicines, to minimise environmental contamination.

On the crop side, this has to be accompanied by environmentally friendly low-input techniques, such as zero tillage, manuring and composting, integrated pest management (IPM) etc. Precise information about soil nutrient levels, coupled with economic incentives, not only helps to minimise production costs but also helps to maintain the soil nutrient balance as well as a landscape that is aesthetically pleasing.

Low-input farming can have a special attraction to consumers who like to buy "environmentally safe" products under "green" labels and to support eco-farming, organic farming, etc. As for any other food product "green label", food must be subject to clear and strict quality standards and be supervised by reliable quality control systems. Prices of such "organic" food products are between 20 and 50 percent higher than conventionally produced foods, mainly as a result of low volumes and stagnant market share. Producers, who are therefore unable to practice economies of scale, nevertheless have to meet high distribution costs.

9

Integrated Pest Management

Pest management is an ecological matter. The size of a pest population and the damage it inflicts is, to a great extent, a reflection of the design and management of a particular agricultural ecosystem. We humans compete with other organisms for food and fiber from our crops. We wish to secure a maximum amount of the food resource from a given area with minimum input of resources and energy. However, if the agricultural system design and/or management is faulty-making it easy for pests to develop and expand their populations or, conversely, making it difficult for predators and parasites of pests to exist-then we will be expending unnecessary resources for pest management. Therefore, the first step in sustainable and effective pest management is looking at the design of the agricultural ecosystem and considering what ecological concepts can be applied to the design and management of the system to better manage pests and their parasites and predators.

The design and management of our agricultural systems need re-examining. We've come to accept routine use of biological poisons in our food systems as normal. But routine use of synthetic chemicals represents significant energy inputs into the agricultural system, and carries both obvious and hidden costs to the farmer and society. Attempting to implement an ecology-based discipline like IPM in large monocultures, which substitute chemical inputs for ecological design, can be an exercise in futility and inefficiency.

IPM, as it was originally conceived, proposed to manage pests though an understanding of their interactions with other organisms and the environment. Despite some differences in emphasis, most of the definitions for IPM have the following elements in common:

— A conception of a managed resource, such as a cropping system on a farm, as a component of a functioning ecosystem. Actions are taken to restore and enhance natural balances in the system, not to eliminate species. Regular monitoring makes it possible to evaluate the populations of pest and beneficial organisms. The producer can then take steps to enhance natural controls (or at least avoid or limit the disruption of natural controls) of the target pest(s).

— An understanding that the presence of a pest does not necessarily constitute a problem. Before a potentially disruptive control method is employed, appropriate decision-making criteria are used to determine whether or not pest management actions are needed.

— A consideration of all possible pest management options before action is taken.

— A philosophy that IPM strategies integrate a combination of all suitable techniques in as compatible a manner as possible; it is important that one technique not conflict with another.

However, IPM has strayed from its ecological roots. Critics of what might be termed "conventional" IPM note that it has been implemented as Integrated Pesticide Management with an emphasis on using pesticides as a tool of first resort. What has been missing from this approach, which is essentially reactive, is an understanding of the ecological basis of pest infestations. Also missing from the conventional approach are guidelines for ecology-based manipulations of the farm agroecosystem that address the questions:

— Why is the pest there?

— How did it arrive?

— Why doesn't the parasite/predator complex control the pest?

Biointensive IPM incorporates ecological and economic factors into agricultural system design and decision making, and addresses public concerns about environmental quality and food safety. The benefits of implementing biointensive IPM can include reduced chemical input costs, reduced on-farm and off-farm environmental impacts, and more effective and sustainable pest management.

An ecology-based IPM has the potential of decreasing inputs of fuel, machinery, and synthetic chemicals-all of which are energy intensive and increasingly costly in terms of financial and environmental impact. Such reductions will benefit the grower and society. Over-reliance on the use of synthetic pesticides in crop protection programs around the world has resulted in disturbances to the environment, pest resurgence, pest resistance to pesticides, and lethal and sub-lethal effects on nontarget organisms, including humans.

These side effects have raised public concern about the routine use and safety of pesticides. At the same time, population increases are placing ever-greater demands upon the "ecological services"—that is, provision of clean air, water and wildlife habitat-of a landscape dominated by farms. Although some pending legislation has recognised the costs to farmers of providing these ecological services, it's clear that farmers and ranchers will be required to manage their land with greater attention to direct and indirect offfarm impacts of various farming practices on water, soil, and wildlife resources. With this likely future in mind, reducing dependence on chemical pesticides in favor of ecosystem manipulations is a good strategy for farmers.

Consumers Union, a group that has carried out research and advocacy on various pesticide problems for many years, defines biointensive IPM as the highest level of IPM:

> a systems approach to pest management based on an understanding of pest ecology.

It begins with steps to accurately diagnose the nature and source of pest problems, and then relies on a range of preventive tactics and biological controls to keep pest populations within acceptable limits.

Reduced-risk pesticides are used if other tactics have not been adequately effective, as a last resort, and with care to minimise risks."

This "biointensive" approach sounds remarkably like the original concept of IPM. Such a "systems" approach makes sense both intuitively and in practice.

The primary goal of biointensive IPM is to provide guidelines and options for the effective management of pests and beneficial organisms in an ecological context. The flexibility and environmental compatibility of a biointensive IPM strategy make it useful in all types of cropping systems.

Even conventional IPM strategies help to prevent pest problems from developing, and reduce or eliminate the use of chemicals in managing problems that do arise. Results of 18 economic evaluations of conventional IPM on cotton showed a decrease in production costs of 7 percent and an average decrease in pesticide use of 15 percent. Biointensive IPM would likely decrease chemical use and costs even further.

Components of Biointensive IPM

An important difference between conventional and biointensive IPM is that the emphasis of the latter is on proactive measures to redesign the agricultural ecosystem to the disadvantage of a pest and to the advantage of its parasite and predator complex. At the same time, biointensive IPM shares many of the same components as conventional IPM, including monitoring, use of economic thresholds, record keeping, and planning.

Planning of IPM program

Good planning must precede implementation of any IPM program, but is particularly important in a biointensive program. Planning should be done before planting because many pest strategies require steps or inputs, such as beneficial organism habitat management, that must be considered well in advance. Attempting to jump-start an IPM program in the beginning or middle of a cropping season generally does not work.

When planning a biointensive IPM program, some considerations include:

— Options for design changes in the agricultural system (beneficial organism habitat, crop rotations)

— Choice of pest-resistant cultivars

— Technical information needs

— Monitoring options, record keeping, equipment, etc.

The Pest Manager

The pest manager is the most important link in a successful IPM program. The manager must know the biology of the pest and the beneficial organisms associated with the pest, and understand their interactions within the farm environment.

As a detailed knowledge of the pest is developed, weak links in its life cycle become apparent. These weak links are phases of the life cycle when the pest is most susceptible to control measures. The manager must integrate this knowledge with tools and techniques of biointensive IPM to manage not one, but several pests. A more accurate title for the pest manager is "ecosystem doctor," for he or she must pay close attention to the pulse of the managed ecosystem and stay abreast of developments in IPM and crop/pest biology and ecology. In this way, the ecosystem manager can take a proactive approach to managing pests, developing ideas about system manipulations, testing them, and observing the results.

IPM options may be considered proactive or reactive. Proactive options, such as crop rotations and creation of habitat for beneficial organisms, permanently lower the carrying capacity of the farm for the pest. The carrying capacity is determined by factors like food, shelter, natural enemies complex, and weather, which affect the reproduction and survival of a species. Cultural controls are generally considered to be proactive strategies.

The second set of options is more reactive. This simply means that the grower responds to a situation, such as an economically damaging population of pests, with some type of shortterm suppressive action. Reactive methods generally include inundative releases of biological controls, mechanical and physical controls, and chemical controls.

Proactive Strategies

Cultural controls are manipulations of the agroecosystem that make the cropping system less friendly to the establishment and proliferation of pest populations. Although they are designed to have positive effects on farm ecology and pest management, negative impacts may also result, due to variations in weather or changes in crop management.

Maintaining and increasing biological diversity of the farm system is a primary strategy of cultural control. Decreased biodiversity tends to result in agroecosystems that are unstable and prone to recurrent pest outbreaks and many other problems. Systems high in biodiversity tend to be more "dynamically stable"—that is, the variety of organisms provide more checks and balances on each other, which helps prevent one species (i.e., pest species) from overwhelming the system.

There are many ways to manage and increase biodiversity on a farm, both above ground and in the soil. In fact, diversity above ground influences diversity below ground. Research has shown that up to half of a plant's photosynthetic production (carbohydrates) is sent to the roots, and half of that (along with various amino acids and other plant products) leaks out the roots into the surrounding soil, providing a food source for microorganisms. These root exudates vary from plant species to plant species and this variation influences the type of organisms associated with the root exudates.

Factors influencing the health and biodiversity of soils include the amount of soil organic matter; soil pH; nutrient balance; moisture; and parent material of the soil. Healthy soils with a diverse community of organisms support plant health and nutrition better than soils deficient in organic matter and low in species diversity.

Research has shown that excess nutrients (e.g., too much nitrogen) as well as relative nutrient balance (i.e., ratios of nutrients?for example, twice as much calcium as magnesium, compared to equal amounts of both) in soils affect insect pest response to plants. Imbalances in the soil can make a plant more attractive to insect pests, less able to recover from pest damage, or more susceptible to secondary infections by plant pathogens.

Soils rich in organic matter tend to suppress plant pathogens. In addition, it is estimated that 75% of all insect pests spend part of their life cycle in the soil, and many of their natural enemies occur there as well. For example, larvae of one species of blister beetle consume about 43 grasshopper eggs before maturing. Both are found in the soil. Overall, a healthy soil with a diversity of beneficial organisms and high organic matter content helps maintain pest populations below their economic thresholds.

Genetic diversity of a particular crop may be increased by planting more than one cultivar. For example, a recent experiment in China demonstrated that disease-susceptible rice varieties planted in mixtures with resistant varieties had 89% greater yield and a 94% lower incidence of rice blast (a fungus) compared to when they were grown in monoculture. The experiment, which involved five townships in 1998 and ten townships in 1999, was so successful that fungicidal sprays were no longer applied by the end of the two-year program.

Species diversity of the associated plant and animal community can be increased by allowing trees and other native plants to grow in fence rows or along water ways, and by integrating livestock into the farm system. Use of the following cropping schemes are additional ways to increase species diversity. Crop rotations radically alter the environment both above and below ground, usually to the disadvantage of pests of the previous crop. The same crop grown year after year on the same field will inevitably build up populations of organisms that feed on that plant, or, in the case of weeds, have a life cycle similar to that of the crop.

Add to this the disruptive effect of pesticides on species diversity, both above and below ground, and the result is an unstable system in which slight stresses (e.g., new pest variety or drought) can devastate the crop. When making a decision about crop rotation, consider the following questions: Is there an economically sustainable crop that can be rotated into the cropping system? Is it compatible? Important considerations when developing a crop rotation are:

- — What two (or three or several) crops can provide an economic return when considered together as a biological and economic system that includes considerations of sustainable soil management?
- — What are the impacts of this season's cropping practices on subsequent crops?
- — What specialised equipment is necessary for the crops?
- — What markets are available for the rotation crops?

A corn/soybean rotation is one example of rotating compatible economic crops. Corn is a grass; soybean is a leguminous broadleaf. The pest complex of each, including soil organisms, is quite different. Corn rootworm, one of the major pests of corn, is virtually eliminated by using this rotation. Both crops generally provide a reasonable return. Even rotations, however, create selection pressures that will ultimately alter pest genetics. A good example is again the corn rootworm: the corn/bean rotation has apparently selected for a small population that can survive a year of non-corn cropping.

Management factors should also be considered. For example, one crop may provide a lower direct return per acre than the alternate crop, but may also lower management costs for the alternate crop, with a net increase in profit.

Other Cropping Structure Options

Multiple cropping is the sequential production of more than one crop on the same land in one year. Depending on the type of cropping sequence used, multiple cropping can be useful as a weed control measure, particularly when the second crop is interplanted into the first.

Interplanting is seeding or planting a crop into a growing stand, for example overseeding a cover crop into a grain stand. There may be microclimate advantages (e.g., timing, wind protection, and less radical temperature and humidity changes) as well as disadvantages to this strategy. By keeping the soil covered, interplanting may also help protect soil against erosion from wind and rain.

Intercropping is the practice of growing two or more crops in the same, alternate, or paired rows in the same area. This technique is particularly appropriate in vegetable production. The advantage of intercropping is that the increased diversity helps "disguise" crops from insect pests, and if done well, may allow for more efficient utilisation of limited soil and water resources. Disadvantages may relate to ease of managing two different crop species—with potentially different nutrient, water, and light needs, and differences in harvesting time and method—in close proximity to each other.

Strip cropping is the practice of growing two or more crops in different strips across a field wide enough for independent cultivation (e.g., alternating six-row blocks of soybeans and corn or alternating strips of alfalfa and cotton or alfalfa and corn). It is commonly practiced to help reduce soil erosion in hilly areas. Like intercropping, strip cropping increases the diversity of a cropping area, which in turn may help "disguise" the crops from pests. Another advantage to this system is that one of the crops may act as a reservoir and/or food source for beneficial organisms. However, much more research is needed on the complex interactions between various paired crops and their pest/ predator complexes.

The options described above can be integrated with no-till cultivation schemes and all its variations (strip till, ridge till, etc.) as well as with hedgerows and intercrops designed for beneficial organism habitat. With all the cropping and tillage options available, it is possible, with creative and informed management, to evolve a biologically diverse, pestsuppressive farming system appropriate to the unique environment of each farm.

Other Cultural Management Options

Disease-free seed and plants are available from most commercial sources, and are certified as such. Use of disease-free seed and nursery stock is important in preventing the introduction of disease.

Resistant varieties are continually being bred by researchers. Growers can also do their own plant breeding simply by collecting non-hybrid seed from healthy plants in the field. The plants from these seeds will have a good chance of being better suited to the local environment and of being more resistant to insects and diseases. Since natural systems are dynamic rather than static, breeding for resistance must be an ongoing process, especially in the case of plant disease, as the pathogens themselves continue to evolve and become resistant to control measures.

Sanitation involves removing and destroying the overwintering or breeding sites of the pest as well as preventing a new pest from establishing on the farm (e.g., not allowing off-farm soil from farm equipment to spread nematodes or plant pathogens to your land). This strategy has been particularly useful in horticultural and tree-fruit crop situations involving twig and branch pests. If, however, sanitation involves removal of crop residues from the soil surface, the soil is left exposed to erosion by wind and water. As with so many decisions in farming, both the short- and long-term benefits of each action should be considered when tradeoffs like this are involved.

Spacing of plants heavily influences the development of plant diseases and weed problems. The distance between plants and rows, the shape of beds, and the height of plants influence air flow across the crop, which in turn determines how long the leaves remain damp from rain and morning dew. Generally speaking, better air flow will decrease the incidence of plant disease. However, increased air flow through wider spacing will also allow more sunlight to the ground, which may increase weed problems. This is another instance in which detailed knowledge of the crop ecology is necessary to determine the best pest management strategies. How will the crop react to increased spacing between rows and between plants? Will yields drop because of reduced crop density? Can this be offset by reduced pest management costs or fewer losses from disease?

Altered planting dates can at times be used to avoid specific insects, weeds, or diseases. For example, squash bug infestations on cucurbits can be decreased by the delayed planting strategy, i.e., waiting to establish the cucurbit crop until overwintering adult squash bugs have died. To assist with disease management decisions, the Cooperative Extension Service (CES) will often issue warnings of "infection periods" for certain diseases, based upon the weather.

In some cases, the CES also keeps track of "degree days" needed for certain important insect pests to develop. Insects, being coldblooded, will not develop below or above certain threshold temperatures. Calculating accumulated degree days, that is, the number of days above the threshold development temperature for an insect pest, makes the prediction of certain events, such as egg hatch, possible.

Some growers gauge the emergence of insect pests by the flowering of certain non-crop plant species native to the farm. This method uses the "natural degree days" accumulated by plants. For example, a grower might time cabbage planting for three weeks after the Amelanchier species (also known as saskatoon, shadbush, or serviceberry) on their farm are in bloom. This will enable the grower to avoid peak egg-laying time of the cabbage maggot fly, as the egg hatch occurs about the time Amelanchier species are flowering. Using this information, cabbage maggot management efforts could be concentrated during a known time frame when the early instars (the most easily managed stage) are active.

Optimum growing conditions are always important. Plants that grow quickly and are healthy can compete with and resist pests better than slow-growing, weak plants. Too often, plants grown outside their natural ecosystem range must rely on pesticides to overcome conditions and pests to which they are not adapted.

Mulches, living or non-living, are useful for suppression of weeds, insect pests, and some plant diseases. Hay and straw, for example, provide habitat for spiders. Research in Tennessee showed a 70% reduction in damage to vegetables by insect pests when hay or straw was used as mulch. The difference was due to spiders, which find mulch more habitable than bare ground. Other researchers have found that living mulches of various clovers reduce insect pest damage to vegetables and orchard crops. Again, this reduction is due to natural predators and parasites provided habitat by the clovers. Vetch has been used as both a nitrogen source and as a weed suppressive mulch in tomatoes in Maryland.

Growers must be aware that mulching may also provide a more friendly environment for slugs and snails, which can be particularly damaging at the seedling stage. Mulching helps to minimise the spread of soilborne plant pathogens by preventing their transmission through soil splash. Mulch, if heavy enough, prevents the germination of many annual weed seeds. Winged aphids are repelled by silver- or aluminum-colored mulches. Recent springtime field tests at the Agricultural Research Service in Florence, South Carolina, have indicated that red plastic mulch suppresses root-knot nematode damage in tomatoes by diverting resources away from the roots (and nematodes) and into foliage and fruit.

Biotech Crops. Gene transfer technology is being used by several companies to develop cultivars resistant to insects, diseases, and herbicides. An example is the incorporation of genetic material from Bacillus thuringiensis (Bt), a naturally occurring bacterium, into cotton, corn, and potatoes, to make the plant tissues toxic to bollworm, earworm, and potato beetle larvae, respectively.

Whether or not this technology should be adopted is the subject of much debate. Opponents are concerned that by introducing Bt genes into plants, selection pressure for resistance to the Bt toxin will intensify and a valuable biological control tool will be lost. There are also concerns about possible impacts of genetically-modified plant products on non-target organisms as well as fears of altered genes being transferred to weed relatives of crop plants. Whether there is a market for gene-altered crops is also a consideration for farmers and processors. Proponents of this technology argue that use of such crops decreases the need to use toxic chemical pesticides.

Biological Control

Biological control is the use of living organisms—parasites, predators, or pathogens—to maintain pest populations below economically damaging levels, and may be either natural or applied. A first step in setting up a biointensive IPM program is to assess the populations of beneficials and their interactions within the local ecosystem. This will help to determine the potential role of natural enemies in the managed agricultural ecosystem. It should be noted that some groups of beneficials (e.g., spiders, ground beetles, bats) may be absent or scarce on some farms because of lack of habitat.

These organisms might make significant contributions to pest management if provided with adequate habitat. Natural biological control results when naturally occurring enemies maintain pests at a lower level than would occur without them, and is generally characteristic of biodiverse systems.

Mammals, birds, bats, insects, fungi, bacteria, and viruses all have a role to play as predators and parasites in an agricultural system. By their very nature, pesticides decrease the biodiversity of a system, creating the potential for instability and future problems. Pesticides, whether synthetically or botanically derived, are powerful tools and should be used with caution.

Creation of habitat to enhance the chances for survival and reproduction of beneficial organisms is a concept included in the definition of natural biocontrol. Farmscaping is a term coined to describe such efforts on farms.

Habitat enhancement for beneficial insects, for example, focuses on the establishment of flowering annual or perennial plants that provide pollen and nectar needed during certain parts of the insect life cycle. Other habitat features provided by farmscaping include water, alternative prey, perching sites, overwintering sites, and wind protection.

Beneficial insects and other beneficial organisms should be viewed as mini-livestock, with specific habitat and food needs to be included in farm planning.

The success of such efforts depends on knowledge of the pests and beneficial organisms within the cropping system. Where do the pests and beneficials overwinter?

What plants are hosts and non-hosts? When this kind of knowledge informs planning, the ecological balance can be manipulated in favor of beneficials and against the pests. It should be kept in mind that ecosystem manipulation is a two-edged sword. Some plant pests (such as the tarnished plant bug and lygus bug) are attracted to the same plants that attract beneficials. The development of beneficial habitats with a mix of plants that flower throughout the year can help prevent such pests from migrating en masse from farmscaped plants to crop plants.

Applied biological control, also known as augmentative biocontrol, involves supplementation of beneficial organism populations, for example through periodic releases of parasites, predators, or pathogens. This can be effective in many situations-well-timed inundative releases of Trichogramma egg wasps for codling moth control, for instance.

Most of the beneficial organisms used in applied biological control today are insect parasites and predators. They control a wide range of pests from caterpillars to mites. Some species of biocontrol organisms, such as Eretmocerus californicus, a parasitic wasp, are specific to one host-in this case the sweetpotato whitefly. Others, such as green lacewings, are generalists and will attack many species of aphids and whiteflies.

Information about rates and timing of release are available from suppliers of beneficial organisms. It is important to remember that released insects are mobile; they are likely to leave a site if the habitat is not conducive to their survival. Food, nectar, and pollen sources can be "farmscaped" to provide suitable habitat. The quality of commercially available applied biocontrols is another important consideration. For example, if the organisms are not properly labeled on the outside packaging, they may be mishandled during transport, resulting in the death of the organisms. A recent study by Rutgers University noted that only two of six suppliers of beneficial nematodes sent the expected numbers of organisms, and only one supplier out of the six provided information on how to assess product viability.

While augmentative biocontrols can be applied with relative ease on small farms and in gardens, applying some types of biocontrols evenly over large farms has been problematic. New mechanised methods that may improve the economics and practicality of large-scale augmentative biocontrol include ground application with "biosprayers" and aerial delivery using small-scale (radio-controlled) or conventional aircraft.

Inundative releases of beneficials into greenhouses can be particularly effective. In the controlled environment of a greenhouse, pest infestations can be devastating; there are no natural controls in place to suppress pest populations once an infestation begins. For this reason, monitoring is very important. If an infestation occurs, it can spread quickly if not detected early and managed. Once introduced, biological control agents cannot escape from a greenhouse and are forced to concentrate predation/parasitism on the pest(s) at hand.

An increasing number of commercially available biocontrol products are made up of microorganisms, including fungi, bacteria, nematodes, and viruses.

Mechanical and Physical Controls

Methods included in this category utilise some physical component of the environment, such as temperature, humidity, or light, to the detriment of the pest. Common examples are tillage, flaming, flooding, soil solarisation, and plastic mulches to kill weeds or to prevent weed seed germination.

Heat or steam sterilisation of soil is commonly used in greenhouse operations for control of soil-borne pests. Floating row covers over vegetable crops exclude flea beetles, cucumber beetles, and adults of the onion, carrot, cabbage, and seed corn root maggots. Insect screens are used in greenhouses to prevent aphids, thrips, mites, and other pests from entering ventilation ducts. Large, multi-row vacuum machines have been used for pest management in strawberries and vegetable crops. Cold storage reduces post-harvest disease problems on produce.

Although generally used in small or localised situations, some methods of mechanical/ physical control are finding wider acceptance because they are generally more friendly to the environment.

Pest Identification

A crucial step in any IPM program is to identify the pest. The effectiveness of both proactive and reactive pest management measures depend on correct identification. Misidentification of the pest may be worse than useless; it may actually be harmful and cost time and money. Help with positive identification of pests may be obtained from university person-nel, private consultants, the Cooperative Extension Service, and books and websites listed under Useful Resources at the end of this publication.

After a pest is identified, appropriate and effective management depends on knowing answers to a number of questions. These may include:

— What plants are hosts and non-hosts of this pest?

— When does the pest emerge or first appear? ?Where does it lay its eggs? In the case of weeds, where is the seed source? For plant pathogens, where is the source(s) of inoculum?

— Where, how, and in what form does the pest overwinter?

— How might the cropping system be altered to make life more difficult for the pest and easier for its natural controls?

Monitoring and economic injury and action levels are used to help answer these and additional questions.

Monitoring

Monitoring involves systematically checking crop fields for pests and beneficials, at regular intervals and at critical times, to gather information about the crop, pests, and natural enemies. Sweep nets, sticky traps, and pheromone traps can be used to collect insects for both identification and population density information. Leaf counts are one method for recording plant growth stages. Square-foot or larger grids laid out in a field can provide a basis for comparative weed counts. Records of rainfall and temperature are sometimes used to predict the likelihood of disease infections. Specific scouting methods have been developed for many crops. The Cooperative Extension Service can provide a list of IPM manuals available in each state. Many resources are now available via Internet.

The more often a crop is monitored, the more information the grower has about what is happening in the fields. Monitoring activity should be balanced against its costs. Frequency may vary with temperature, crop, growth phase of the crop, and pest populations. If a pest population is approaching economically damaging levels, the grower will want to monitor more frequently.

Economic Injury and Action Levels

The economic injury level (EIL) is the pest population that inflicts crop damage greater than the cost of control measures. Because growers will generally want to act before a population reaches EIL, IPM programs use the concept of an economic threshold level (ETL or ET), also known as an action threshold. The ETL is closely related to the EIL, and is the point at which suppression tactics should be applied in order to prevent pest populations from increasing to injurious levels.

In practice, many crops have no established EILs or ETLs, or the EILs that have been developed may be static over the course of a season and thus not reflect the changing nature of the agricultural ecosystem. For example, a singlecutworm can do more damage to an emerging cotton plant than to a plant that is six weeks old. Clearly, this pest's EIL will change as the cotton crop develops.

ETLs are intimately related to the value of the crop and the part of the crop being attacked. For example, a pest that attacks the fruit or vegetable will have a much lower ETL (that is, the pest must be controlled at lower populations) than a pest that attacks a non-saleable part of the plant. The exception to this rule is an insect or nematode pest that is also a disease vector. Depending on the severity of the disease, the grower may face a situation where the ETL for a particular pest is zero, i.e., the crop cannot tolerate the presence of a single pest of that particular species because the disease it transmits is so destructive.

Special Considerations

Cosmetic Damage and Aesthetics

Consumer attitudes toward how produce looks is often a major factor when determining a crop's sale price. Cosmetic damage is an important factor when calculating the EIL, since pest damage, however superficial, lowers a crop's market value. Growers selling to a market that is informed about IPM or about organically grown produce may be able to tolerate higher levels of cosmetic damage to their produce.

Record-keeping

Monitoring goes hand-in-hand with recordkeeping, which forms the collective "memory" of the farm. Records should not only provide information about when and where pest problems have occurred, but should also incorporate information about cultural practices (irrigation, cultivation, fertilisation, mowing, etc.) and their effect on pest and beneficial populations. The effects of nonbiotic factors, especially weather, on pest and beneficial populations should also be noted. Record-keeping is simply a systematic approach to learning from experience. A variety of software programs are now available to help growers keep track of-and access-data on their farm's inputs and outputs.

Time and Resources

A successful biointensive IPM program takes time, money, patience, short- and long-term planning, flexibility, and commitment. The pest manager must spend time on self-education and on making contacts with Extension and research personnel. Be aware that some IPM strategies, such as increasing beneficial insect habitat, may take more than a year to show results.

A well-run biointensive IPM system may require a larger initial outlay in terms of time and money than a conventional IPM program. In the long run, however, a good biointensive IPM program should pay for itself. Direct pesticide application costs are saved and equipment wear and tear may be reduced.

Chemical Controls

Synthetic pesticides comprise a wide range of man-made chemicals used to control insects, mites, weeds, nematodes, plant diseases, and vertebrate and invertebrate pests. These powerful chemicals are fast acting and relatively inexpensive to purchase.

Pesticides are the option of last resort in IPM programs because of their potential negative impacts on the environment, which result from the manufacturing process as well as from their application on the farm.

Pesticides should be used only when other measures, such as biological or cultural controls, have failed to keep pest populations from approaching economically damaging levels.

If chemical pesticides must be used, it is to the grower's advantage to choose the least-toxic pesticide that will control the pest but not harm non-target organisms such as birds, fish, and mammals. Pesticides that are short-lived or act on one or a few specific organisms are in this class. Examples include insecticidal soaps, horticultural oils, copper compounds (e.g., bordeaux mix), sulfur, boric acid, and sugar esters.

Biorational pesticides

Although use of this term is relatively common, there is no legally accepted definition. Biorational pesticides are generally considered to be derived from naturally occurring compounds or are formulations of microorganisms. Biorationals have a narrow target range and are environmentally benign. Formulations of Bacillus thuringiensis, commonly known as Bt, are perhaps the bestknown biorational pesticide. Other examples include silica aerogels, insect growth regulators, and particle film barriers.

Particle film barriers

A relatively new technology, particle film barriers are currently available under the tradename Surround??WP Crop Protectant. The active ingredient is kaolin clay, an edible mineral long used as an anti-caking agent in processed foods, and in such products as toothpaste and Kaopectactate. There appears to be no mammalian toxicity or any danger to the environment posed by the use of kaolin in pest control. The kaolin in Surround is processed to a specific particle size range, and combined

cost the grower money and increase the risk of environmental damage. If the monitoring program indicates that the pest outbreak is isolated to a particular location, spot treatment of only the infested area will not only save time and money, but will conserve natural enemies located in other parts of the field. The grower should also time treatments to be least disruptive of other organisms. This is yet another example where knowledge about the agroecosystem is important.

With the increasing popularity of no-till and related conservation tillage practices, herbicide use has increased. One way to increase application efficiency and decrease costs of herbicide use is through band application. This puts the herbicide only where it is needed, usually in soil disturbed by tillage or seed planting, where weeds are most likely to sprout.

Baits and microencapsulation of pesticides are promising technologies. For example, Slam? is an insecticide-bait mixture for control of corn rootworm. It is a formulation of a bait, curcubitacin B, and carbaryl in microspheres. It is selective, and reduces the amount of carbaryl needed to control the rootworm by up to 90%.

Another example of bait-insecticide technology is the boll weevil bait tube. It lures the boll weevil using a synthetic sex pheromone. Each tube contains about 20 grams of malathion, which kills the boll weevil. This technique reduces the pesticide used in cotton fields by up to 80% and conserves beneficials. It is most effective in managing low, earlyseason populations of the boll weevil.

Integrated Weed Management Systems

Weeds as competitors in crops present a number of unique challenges that need to be recognised when developing management strategies. The intensity of weed problems during a growing season will be influenced by weed population levels in previous years. The axiom "one year's seeding equals seven years' weeding" is apt.

Weed control costs cannot necessarily be calculated against the current year's crop production costs. Weeds present a physical problem for harvesting. Noxious weed seed mixed with grain reduces the price paid to growers. If the seed is sold for crop production the weed can be spread to new areas. For example, the perennial pepperweed, thought to have been introduced to California in sugar beet seed, now infests thousands of acres in the state. In addition, weed economic thresholds must take into account multiple species and variable competetive ability of different crops. For example, 12.7 cocklebur plants in 10 sq. meters of corn cause a 10% yield loss. Only 2 cockleburs in the same area planted to soybeans will cause the same 10% crop loss.

Tactics that can be integrated into weed management systems include:

— *Prevention* - The backbone of any successful weed management strategy is prevention. It is important to prevent the introduction of seeds into the field through sources like irrigation water or manure.

— *Crop rotation* -A practical and effective method of weed management.

— *Cultivation* - Steel in the Field: A Farmer's Guide to Weed Management Tools shows how today's implements and techniques can handle weeds while reducing or eliminating herbicides.

— *Flame weeding* - good for control of small weeds.

— *Delayed planting* - Early-germinating weeds can be destroyed by tillage. And with warmer weather, the subsequently planted crop (depending on the crop, of course) will grow more quickly, thus competing better with weeds.

— *Staggered planting schedule* - This will allow more time for mechanical weed control, if needed. This also lessens the weather risks and spaces out the work load at harvest time.

— *Surface residue management* - A thick mulch may shade the soil enough to keep weed seeds from germinating. In addition, some plant residues are allelopathic, releasing compounds that naturally suppress seed germination.

— *Altered plant spacing or row width* - An example is narrow-row soybean plantings. The faster the leaves shade the ground, the less weeds will be a problem.

— *Herbivores* - Cattle, geese, goats, and insects can be used to reduce populations of specific weeds in special situations. Cattle, for example, relish Johnson grass. Weeder geese were commonly used in cotton fields before the advent of herbicides. Musk thistle populations can be satisfactorily reduced by crown- and seedeating weevils. Goats may be used for large stands of various noxious weeds.

— *Adjusting herbicide use to situation* - Herbicide selection and rate can be adjusted depending upon weed size, weed species, and soil moisture. Young weeds are more susceptible to chemicals than older weeds.

By integrating a variety of tactics, farmers can reduce or eliminate herbicide use.

Current Status of IPM

In the last twenty years or so, IPM programs have been developed for important pests in corn, soybeans, cotton, citrus, apples, grapes, walnuts, strawberries, alfalfa, pecans, and most other major crops. These programs are constantly being revised or fine-tuned, and occasionally undergo a significant overhaul as the introduction of a new technology or new pest makes the present IPM program obsolete.

The primary goal of biointensive IPM is to provide guidelines and options for the effective management of pests and beneficial organisms in an ecological context. This requires a somewhat different set of knowledge from that which supports conventional IPM, which in turn requires a shift in research focus and approach. Recommended actions to better facilitate the transition to biointensive IPM are:

— Build the knowledge/information infrastructure by making changes in research and education priorities in order to emphasise ecology-based pest management

— Redesign government programs to promote biointensive IPM, not "Integrated Pesticide Management"

— Offer consumers more choices in the marketplace

— Use the market clout of government and large corporations

— Use regulation more consciously, intelligently, and efficiently

IPM in the future will emphasise biological and ecological knowledge in managing pests. Beyond that, specific areas are described here that will impact research and implementation of IPM in the future.

Pest control methods are evolving and diversifying in response to public awareness of environmental and health impacts of synthetic chemical pesticides and resulting legislation.

The strong growth of the organic foods market—20% annual expansion for the past several years—may also be a factor in the accelerated development of organic pest management methods.

Agricultural pests are developing resistance to many synthetic agrichemicals, and new synthetic chemicals are being registered at a slower rate than in the past. This situation has helped open the market for a new generation of microbial pesticides.

Research is proceeding on natural endophytes - fungi or bacteria that have a symbiotic (mutually beneficial) relationship with their host plant-and their effects on plant pests. This research might yield products that could be used to inoculate plants against certain pests.

More Weed IPM

Weeds are the major deterrent to the development of more sustainable agricultural systems, particularly in agronomic crops. Problems associated with soil erosion and water quality are generally the result of weed control measures like tillage, herbicides, cultivation, planting date and pattern, etc.. In the future, research will focus not on symptoms, such as soil erosion, but on basic problems such as how to sustainably manage soils.

On-farm Resources

As farm management strategies become increasingly fine-tuned to preserve a profitable bottom line, the conservation, utilisation, and development of on-farm resources will take on added importance. In the context of

IPM, this will mean greater emphasis on soil management as well as on conserving beneficial organisms, retaining and developing beneficial habitats, and perhaps developing on-farm insectaries for rearing beneficial insects.

IPM On-line

There is an increasing body of information about production, marketing, and recordkeeping available to growers via the Internet. The Internet is also a good source of information about IPM, beneficial insects, products, and pest control options for individual crops. IPM specialists are generating highquality websites as a modern educational delivery tool, and many Extension Service leaflets are now being made available in electronic format only. This trend will only accelerate as more and more agriculturists familiarise them-selves with the Internet.

IPM Certification and Marketing

Certification of crops raised according to IPM or some other ecology-based standards may give growers a marketing advantage as public concerns about health and environmental safety increase. For example, since 1995, Wegmans has sold IPM-labeled freshmarket sweet corn in its Corning, Geneva, Ithaca, Syracuse, and Rochester, New York stores. Wegmans has also added IPM-labeled corn, beets, and beans to its shelves of canned vegetables. One goal of the program, in addition to being a marketing vehicle, is to educate consumers about agriculture and the food system. Another goal is to keep all growers moving along the "IPM Continuum."

Ecolabels are becoming more popular, with over a dozen brands now in existence. They may provide for a more certain market and perhaps a price premium to help growers offset any costs associated with implementing sustainable farming practices. A possible downside to implementing such programs is that they require additional paperwork, development of standards and guidelines, and inspections. There is concern from some quarters that IPM labeling will cause consumers to raise more questions about pesticide use and the safety of conventional produce. Some advocates of organic farming worry about consumer confusion over the relationship of the ecolabel to the "Certified Organic" label.

Bibliography

Allen, C.L., and M.H. Entz, "Zero-tillage establishment of alfalfa and meadow bromegrass as influenced by previous annual grain crop", *Can. J. Plant Sci.*, 1994.

Badaruddin, M., and D.W. Meyer, "Legume forages effects on soil nitrogen and grain yield, and nitrogen nutrition of wheat", Agron. J., 1989.

Carr, P.M., B.G. Schatz, J.C. Gardner, and S.F. Zwinger, "Grain yield and biomass from intercropping wheat and flax". J. Prod. Agric, 1993.

______________., R.D. Horsley, and W.W. Poland, "Tillage and seeding rate effects on wheat cultivars: I", *Grain production*, Crop Sci., 2003.

Holford, I.C.R., "Effects of duration of grazed lucerne on long-term yields and nitrogen uptake of subsequent wheat", Aust. J. Agric. Res., 1980.

______________., "Yields and nitrogen uptake of grain sorghum in various rotations including lucerne, annual legume and long fallow", Aust. J. Agric, 1989.

Littler, J.W., "Effect of pasture on subsequent wheat crops on a black earth soil of the Darling Downs", *1. The overall experiment*, Qld. J. Agric. & Animal Sciences., 1984.

Miller, P.R., J. Waddington, C.L. McDonald, and D.A. Derksen, "Cropping sequence affects wheat productivity on the semiarid northern Great Plains", *Can. J. Plant Sci.*, 2002.

Schiere, J.B. and De Wit, J., "Livestock and farming systems research II:development and classifications", pp. 39 - 61 in

J.B.Schiere, *Cattle, straw and systems control,* Royal Tropical Institute, Amsterdam, 1995.

Schiere, J.B., G.J. Oomen and M.N.M. Ibrahim, "Mixed crop-livestock farming: different systems and their trade-offs", *Invited paper at the annual workshop of the South African Society of Animal Sciences,* University of Zululand, Faculty of Agriculture, Empangeni, South Africa, October 1997.

Shewmaker, G.E., H.F. Mayland and S.B. Hansen, "Cattle grazing preference among eight endophyte-free tall fescue cultivars", Agron. J., 1997.

Thomas G.A. et al., "Nitrogen fertilisation of wheat under different fallow management practices in SW Queensland", In: Proc of 6th Aust. Agron. *Conference,* Armidale, 1992.

Wylie P.B., "Integrated Farming Systems", *In:Sustainable Agriculture in the Sub-tropics, an Australian perspective,* Eds: Clarke A.L. and Wylie P.B. DPI, Brisbane, 1997.

Index